Mechanical engineering science

Mechanical engineering science

G. D. JONES B.Sc (Eng), C.Eng, MI MechE
Principal Lecturer and Director of Studies
School of Engineering
College of Further Education, Plymouth

Longman
Scientific &
Technical

Longman Scientific & Technical,
Longman Group UK Limited,
Longman House, Burnt Mill, Harlow,
Essex CM20 2JE, England
and Associated Companies throughout the world.

First published 1989

British Library Cataloguing in Publication Data
Jones, G. D.
  Mechanical engineering science.
  1. Mechanical engineering
  I. Title
  621

ISBN 0-582-00944-8

Set in Compugraphic Times

Produced by Longman Group (FE) Ltd
Printed in Hong Kong

Contents

Introduction

Introduction

The aim of this book is to cover the Mechanical Science requirements of Technicians and trainee engineers up to first year HNC/HND and Degree level. It will be particularly useful for the final years of BTEC National Certificate courses in Engineering, the first years of BTEC Higher National Certificate and Diploma courses in Engineering, and for GCE 'A'-level students embarking on HND and Degree courses at Polytechnic or University.

I have attempted to write the text logically in small steps and with numerous worked examples so that it could easily form the basis of a self-study text. In this respect I hope that the book will be suitable for students studying without the benefit of constant tutor support, as may be the case on distance-learning schemes for example.

With the current vogue towards resource-based learning and a more student-centred approach to teaching it is important that teachers spend time in devising and developing suitable practical activities to augment the theoretical treatment of the subject that this book promotes. It is only by practical application that the use of the higher-order skills of analysis, synthesis and evaluation will be developed in technicians at these higher levels.

Finally, if the book falls short of the ideal that I had in mind, then I can only hope that it provokes its readers to recognise and rectify its faults and deficiencies. As always, I am indebted to all those colleagues who aided me in this task, and my wife, Lynne, for typing the manuscript.

G. D. JONES
Plymouth 1989

1 Basic mechanics

The contents of this chapter enable you to solve problems involving momentum, energy and power.

1.1 Momentum

The linear momentum of a body is defined as the product of the mass of the body and its velocity. Since velocity is a vector quantity, momentum too is a vector quantity, so that

$$\text{Linear momentum} = mv \tag{1.1}$$

and its units are kg × m/s or Ns since 1 N = 1 kgm/s^2.

The momentum of a body cannot be altered unless acted upon by an external force for a certain length of time. A force acting for a short period of time is called an **impulse**. Thus the **law of conservation of momentum** states that the total momentum of a body, or system, remains unchanged unless acted upon by an external impulse.

For example, consider a block of wood of mass M hanging from a string (Fig. 1.1). A bullet fired into the block from a gun will possess a certain momentum equal to the mass of the bullet (m) multiplied by its velocity, V. On impact the block will instantaneously begin to move in the same direction as the bullet with a velocity of its own, v, which will also be the new velocity of the bullet (assuming the bullet has embedded itself in the block). Since the system is not acted upon by an external impulse, according to the law of conservation of momentum

$$\text{momentum before impact} = \text{momentum after impact} \tag{1.2}$$

$$\text{i.e. } mV = (m + M)v$$

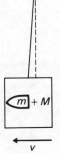

Fig. 1.1

Example 1.1

An engine of mass 10 000 kg travelling at 10 km/h collides with a train of goods wagons of total mass 8000 kg. Assuming that there is no rebound on impact, calculate the common velocity of the train after the collision.

Let v m/s be the common velocity of the train. The momentum before impact is only that of the engine, since the wagons are not moving.

$$10 \text{ km/h} = \frac{10 \times 10^3}{60 \times 60} = 2 \cdot 78 \text{ m/s}$$

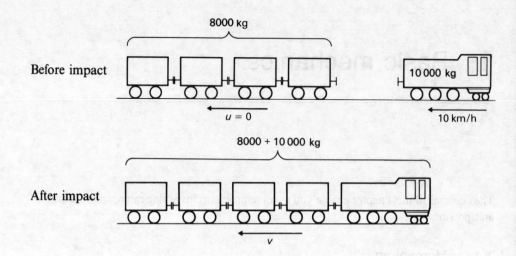

therefore momentum before $= mV = 10\ 000 \times 2 \cdot 78 = 27\ 800$ Ns.

The momentum after impact is the product of the combined mass of engine and wagons and their common speed

therefore momentum after $= Mv = (10\ 000 + 8000)v = 18\ 000v$

Using the conservation of momentum,

momentum before $=$ momentum after

$$27\ 800 = 18\ 000v$$

therefore $v = \dfrac{27\ 800}{18\ 000} = 1 \cdot 54$ m/s

Common velocity of train after impact is $1 \cdot 5$ m/s or $5 \cdot 4$ km/h.

1.2 Newton's second law of motion

This law states that if the momentum of a body changes, the *rate* at which it changes is proportional to the net force causing the change, i.e.

$$F = k\,\frac{d(mv)}{dt}$$

In SI the units of force and momentum are such that the constant k is unity, so that

$$F = \frac{d(mv)}{dt} \qquad\qquad (1.3)$$

In many examples the mass will be constant, and so the equation can be written

$$F = m\,\frac{dv}{dt}$$

but $\dfrac{dv}{dt} = a$, acceleration therefore

$$F = ma \qquad (1.4)$$

That is, if a net out-of-balance force of F newtons acts on a body of mass m kg it will cause the body to accelerate at a m/s^2.

1.3   Impulse

As described in section 1.1, impulse is the product of a force and the time for which it acts. Thus for a constant force F acting for a short time t

$$\text{Impulse} = Ft \quad \text{Ns} \qquad (1.5)$$

If an impulse is applied to a body, the result will be a change in the momentum of the body. Thus

impulse = change in momentum,

i.e. $Ft = mv_2 - mv_1$

or $Ft = m(v_2 - v_1) \qquad (1.6)$

(Remember that force = *rate* of change of momentum)

Example 1.2

A hammer of mass 14 kg hits a wedge with a velocity of 15 m/s and rebounds with a velocity of 2 m/s. The duration of the impact is $0\cdot01$ seconds. Calculate the average force exerted on the wedge.

impulse = change in momentum

Let F be the unknown force, the initial velocity is 15 m/s, the final velocity is -2 m/s (taking downwards direction as positive).

Therefore $F \times 0\cdot01 = 14[15 - (-2)]$

therefore $F = \dfrac{14 \times 17}{0\cdot01} = 23\ 800$ N

$$= 23\cdot8 \text{ kN}$$

Problems 1.1

1   A ball of mass 110 gm is projected along a smooth hard floor at 8 m/s perpendicular to a wall which it strikes and rebounds with a velocity of 7 m/s. If the time of contact with the wall is $\frac{1}{200}$ th s, find the impact force exerted by the ball on the wall.

2   In a drop stamping press the die, which has a mass of 25 kg, is dropped through a height of $5\cdot3$ m and after impact with the work on the anvil rises to a height of $0\cdot75$ m.
   a   Calculate the velocity of the die immediately before and immediately after impact.
   b   If the die is in contact with the work for $\frac{1}{20}$ th s calculate the impact force of the blow.

3 A cricket ball, of mass 156 gm, reaches a batsman with a velocity of 7 m/s and is driven straight back with a velocity of 20 m/s. Calculate the force between the bat and the ball if they are in contact for $\frac{1}{10}$ th s.

4 A and B are two wagons moving in the same straight line. A has a mass of 4 Mg and moves at 5·3 m/s to the right. B has a mass of 8 Mg and moves at 0·6 m/s to the left. If the two wagons collide and then begin to move with a common velocity determine:
  a the magnitude and direction of the common velocity
  b the distance the two wagons move together after impact if there is a resistance to motion of 62 N/Mg.

5 **a** State the principle known as the principle of conservation of linear momentum.
  b A bullet of mass 45 g is fired into a freely suspended target of mass 5 kg. On impact the target moves with an initial velocity of 7 m/s.
    i Calculate the velocity of the bullet.
    ii If the penetration is 150 mm, find the average force exerted.

6 A naval gun barrel and carriage have a mass of 40 Mg. The gun fires a shell of mass 80 kg with a velocity of 350 m/s. The recoil cylinders exert a force of 665 kN. Calculate:
  a the velocity with which the gun begins to recoil
  b the distance travelled on the recoil
  c the time taken on the recoil.

7 A truck of mass 10 Mg and travelling at 10 km/h collides with a stationary truck of mass 5 Mg. After the collision the latter travels 140 m before being brought to rest against an average retarding force of 66 N. Calculate the velocity of each truck immediately after the impact.

8 A pile-driver of mass 110 kg falls from rest a vertical distance of 5·3 m and strikes a pile of mass 560 kg which it drives a distance of 75 mm into the ground. Assume that after the blow the pile-driver and pile move together. Calculate:
  a the velocity of the pile-driver immediately before impact
  b the velocity of the pile-driver and pile immediately after impact
  c the resistance, assumed uniform, which the ground offers.

9 A billiard ball A strikes another billiard ball B which is travelling in the opposite direction at 6 m/s. What would have to be the impact velocity of A if B remains stationary after the impact and A rebounds with a velocity of 4 m/s? Both balls are of mass 400 g. If the impact force was 2400 N what was the time of impact?

10 A truck of mass 6 Mg travelling at 8·8 m/s collides with another truck of mass 4 Mg travelling at 4·5 m/s in the same direction. The impact lasts for 0·5 seconds after which the trucks are travelling together at the same speed. Calculate:
  a the speed of the trucks resulting from the collision
  b the force acting on each truck during the collision.

1.4 The impulse of a variable force

The concept of impulse is particularly important when the force is variable. The impulse of a variable force is the product of the magnitude of the force at a particular instant and the length of time for which the force maintains that value. For a force which is varying with time it will usually be necessary to know the relationship between the two variables, i.e. F as a function of time. If this is so, using equation (1.3),

$$F = \frac{d}{dt}(mv)$$

Integrating both sides of the equation with respect to time between the limits t_1 and t_2,

$$\int_{t_1}^{t_2} F\,dt = m \int_{v_1}^{v_2} dv \quad \text{assuming } m \text{ is constant}$$

$$= m[v]_{v_1}^{v_2}$$

$$= m(v_2 - v_1) \tag{1.7}$$

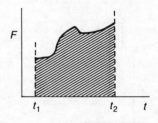

Now $\int_{t_1}^{t_2} F\,dt$ is the area under the graph of F against t, and $m(v_2 - v_1)$ is the change in momentum.

Thus, the change in momentum of a body subject to a force varying with time is equal to the area under the force-time graph which in turn is the **total impulse** given to the body.

Example 1.3

A force exerted on a body of mass 10 kg travelling with an initial velocity 15 m/s varies with time according to the relationship $F = 250 + 2t^2$. Calculate the total impulse imparted to the body in the first 3 seconds, and the final velocity of the body.

$$\text{Total impulse} = \int_{t_1}^{t_2} F\,dt = \int_0^3 (250 + 2t^2)\,dt$$

$$= \left[250t + \frac{2t^3}{3}\right]_0^3$$

$$= [750 + 18] - [0]$$

$$= 768 \text{ Nm}$$

$$\text{but total impulse} = \text{change in momentum of body}$$

$$\text{therefore } 768 = m(v_2 - v_1)$$

$$= 10(v_2 - 15)$$

$$\text{therefore } v_2 = \frac{768}{10} + 15$$

$$= 91 \cdot 8 \text{ m/s}$$

1.5 Work and energy

Energy is the capacity to do work. In mechanical problems work is done on a body in changing its position or its velocity. When a constant force F moves a body through a distance *s*, the amount of work done is equal to the product of the force and the distance moved in the direction of the force, i.e.

$$\text{work done in moving body} = Fs \text{ joules} \tag{1.8}$$

Similarly, work is done *by* a body in moving a distance *s* *against* a force *F* thus:

$$\text{work done by body against a force} = Fs \text{ joules}$$

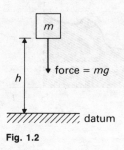

Fig. 1.2

POTENTIAL ENERGY

In the absence of an external force, the energy required to do this amount of work is called the **potential energy** of the body. This energy is described as the energy possessed by a body due to the position or height of the body above some datum.

For example, consider a mass *m* (kg) at a height *h* (m) above a datum (Fig. 1.2). The potential energy possessed by the mass is equal to the work that would be done in falling down to the datum, i.e.

$$\text{work done} = \text{force on body (due to gravity)} \times \text{distance moved}$$
$$= mgh \tag{1.9}$$

Therefore potential energy possessed by a body at a height *h* above a datum is *mgh*.

The converse of the above analysis is that the work done in *raising* a body of mass *m* to a height *h* is *mgh* and is equal to the amount of potential energy given to, and stored by, the body.

Example 1.4

A mass of 8 kg is held at a height of 2 m above the ground. Calculate the potential energy of the mass. If the mass is now allowed to fall to the ground calculate the work done.

$$\text{Potential energy of body} = mgh$$
$$= 8 \times 9 \cdot 81 \times 2$$
$$= 157 \text{ joules}$$

$$\text{Work done} = \text{loss in potential energy}$$
$$= 157 \text{ joules}$$

KINETIC ENERGY

A constant out-of-balance force *F* acting on a body of mass *m* will cause a change in velocity of the body, i.e. an acceleration *a* which will also be constant. The relationship is given by Newton's second law:

$$\text{(from 1.4)} \quad F = ma$$

Consider a body under the action of such a constant force. Suppose the body attained a velocity *v* from rest in time *t* and in a distance *s*, then the work done by the body = *Fs* joules.

Since the acceleration is constant, $v^2 = u^2 + 2as$ and $u = 0$, therefore

$$s = \frac{v^2}{2a}$$

Now, work done

$$= Fs$$

$$= (ma)\left(\frac{v^2}{2a}\right)$$

$$= \tfrac{1}{2}mv^2 \tag{1.10}$$

The expression $\tfrac{1}{2}mv^2$ is termed the **kinetic energy** of the body, which is described as the energy possessed by a body due to its motion or velocity.

Thus the work done on a body in giving it a velocity of v m/s from rest is equal to the kinetic energy of the body ($\tfrac{1}{2}mv^2$).

A graph of force F, against distance moved from S_1 to S_2 is given in Fig. 1.3. The area under this graph from S_1 to S_2 is given by $\int_{S_1}^{S_2} F \, ds$ or $F(S_2 - S_1)$, which is clearly the area of the shaded rectangle in Fig. 1.3.

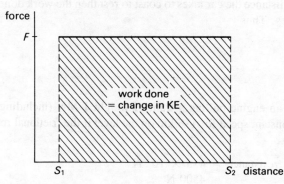

Fig. 1.3

Thus the work done by a force F in moving a body from S_1 to S_2 is equal to the area under the force-distance graph and is also equal to the change in kinetic energy of the body. If v_1 and v_2 are the initial and final velocities, then

$$F(S_2 - S_1) = \tfrac{1}{2}m(v_2^2 - v_1^2) \tag{1.11}$$

Example 1.5

A motor car of mass 850 kg travelling at 100 km/h has its brakes applied, which reduces its speed to 25 km/h. Calculate the change in kinetic energy. If the braking force required is constant at 375 N, calculate the distance travelled during the speed reduction.

$$100 \text{ km/h} = \frac{100 \times 10^3}{60 \times 60} \text{ m/s} = 27 \cdot 78 \text{ m/s}$$

and $\quad 25 \text{ km/h} = 6 \cdot 94 \text{ m/s}$

$$\text{change in KE} = \tfrac{1}{2}m(v_2^2 - v_1^2) = \tfrac{1}{2} \times 850 \times (27 \cdot 78^2 - 6 \cdot 94^2)$$

$$= 425(771 \cdot 73 - 48 \cdot 16)$$

$$= 307 \cdot 5 \text{ kJ}$$

Also, change in KE = work done by braking force = Fs

$$\text{therefore } s = \frac{\text{change in KE}}{F} = \frac{307 \cdot 5 \times 10^3}{375} = 820 \text{ m}$$

WORK DONE AGAINST FRICTION

In many practical situations the motion of a body is opposed by resistive forces such as wind resistance, bearing resistance, frictional resistance of rough surfaces etc. These resistances can be regarded as a single force opposing the motion. Thus the body is continually doing work against this frictional force. A body moving at a steady velocity (i.e. not accelerating) is doing enough work to provide a tractive pull equal but opposite to the resistive force. In order to accelerate, the body needs to do enough work to overcome the resistance *and* provide an extra pull for acceleration.

Suppose the driver of a car moving with a steady velocity of v m/s switches off the power and hence the pull of the engine. The kinetic energy possessed by the car ($\frac{1}{2}mv^2$) will be used to do work against the resistive forces. If R is the constant resistive force and s the distance the car takes to coast to rest then the work done against friction will be Rs joules. Thus

$$Rs = \tfrac{1}{2}mv^2 \tag{1.12}$$

Example 1.6

Find the work done by an engine pulling a train of weight 1 MN (including engine weight) travelling at a constant speed for $1 \cdot 6$ km on the level if the frictional resistance is $4 \cdot 5$ N per kN weight.

$$\text{Total frictional resistance} = 4 \cdot 5 \times 10^3 \text{ N}$$
$$= 4500 \text{ N}$$

$$\text{Work done against resistance} = Rs$$
$$= 4500 \times 1 \cdot 6 \times 10^3$$
$$= 7\,200\,000 \text{ J}$$
$$= 7200 \text{ kJ}$$

Example 1.7

A lorry is travelling at 30 km/h up a slope of 1 in 12. The lorry has a mass of 6000 kg and the total frictional force is 310 N. Find the work done per minute to overcome (a) friction and (b) gravity.

$$\text{In 1 min the lorry travels } \frac{30}{60} \text{ km} = 500 \text{ m}$$

therefore work done per minute against friction = $310 \times 500 = 155$ kJ.

For a distance of 500 m along the slope the lorry rises a vertical height of $\frac{500}{12}$ m, therefore work done per min against gravity

$$= mgh = 6000 \times 9 \cdot 81 \times \tfrac{500}{12} = 2452 \cdot 5 \text{ kJ}$$

WORK DONE BY A VARIABLE FORCE

If the force acting to move a body from S_1 to S_2 is variable, the work done is still represented by the area under the force-distance graph from S_1 to S_2 and hence by $\int_{S_1}^{S_2} F\,ds$ (Fig. 1.4). This work done can only be evaluated if the relationship between the

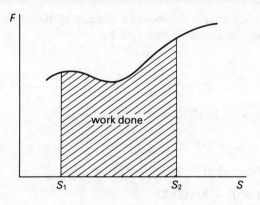

Fig. 1.4

force and some other variable is known. Usually the other variable will be either distance, velocity or time.

FORCE VARYING WITH DISTANCE

Using Newton's second law

$$F = \frac{d}{dt}(mv)$$

or $$F = m\frac{dv}{dt}$$

where F is a function of the distance moved by the body.

Now $$\frac{dv}{dt} = \frac{ds}{dt} \times \frac{dv}{ds} = v\frac{dv}{ds}$$

therefore $$F = mv\frac{dv}{ds}$$

Integrating both sides with respect to S between the limits S_1 and S_2

$$\int_{S_1}^{S_2} F\,ds = \int_{v_1}^{v_2} mv\,dv$$

or $$\int_{S_1}^{S_2} F\,ds = m\left|\frac{v^2}{2}\right|_{v_1}^{v_2}$$

$$= \tfrac{1}{2}m(v_2^2 - v_1^2) \tag{1.13}$$

This again shows that the area under the force-distance graph between S_1 and S_2 is equal to the increase in kinetic energy of the body.

Example 1.8

The force F on a body of mass 50 kg varies with the distance s of the body from a fixed point according to the law

$$F = 202 \cdot 5 - 2s \text{ kN}$$

Calculate the velocity of the body after it has moved a distance of 200 m from rest, and the work done by the force in achieving this velocity.

$$\int_{s_1}^{s_2} F \, ds = \tfrac{1}{2} m(v_2^2 - v_1^2)$$

therefore $\displaystyle\int_0^{200} (202 \cdot 5 - 2s) 10^3 ds = \tfrac{1}{2} \times 50(v_2^2 - 0)$

therefore $10^3 [202 \cdot 5s - s^2]_0^{200} = 25 v_2^2$

$$10^3 \{ [202 \cdot 5 \times 200 - 200^2] - [0] \} = 25 \, v_2^2$$

therefore $v_2^2 = (40\ 500 - 40\ 000)40$

$$v_2 = 141 \cdot 4 \text{ m/s}$$

Work done by force = change in kinetic energy

$$= \tfrac{1}{2} m(v_2^2 - v_1^2)$$
$$= \tfrac{1}{2} \times 50 \times (141 \cdot 4)^2$$
$$= 499 \cdot 85 \text{ kJ}$$

Example 1.9

The force F acting on a body of mass 100 kg varies with the distance S of the body from a fixed point as shown by the following table:

force (kN)	5·8	6·0	6·2	6·8	7·5	8·2	8·8	9·1	9·4	9·5
distance (m)	0	20	30	40	50	60	70	80	90	100

Calculate the total work done in moving the body from rest a distance of 100 m, and the velocity at that distance.

The total work done is equal to the area under the force-distance graph, the work diagram. There are a number of methods of computing the area under the curve, for example using Simpson's rule and dividing the area into 10 strips (12 ordinates) each of width 10 m.

$$\text{Area} = \frac{h}{3} (A + 2B + 4C)$$

where A = sum of first and last ordinate, B = sum of remaining odd ordinates, and C = sum of even ordinates.

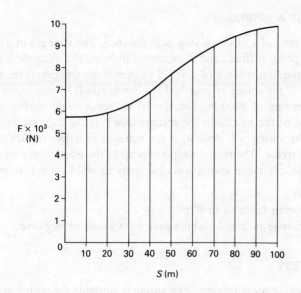

(1) Ordinate	(2) Length	(3) Simpson's multiplier	Product of (2) × (3)
1	5·8	1	5·8
2	5·8	4	23·2
3	6·0	2	12·0
4	6·2	4	24·8
5	6·8	2	13·6
6	7·5	4	30·0
7	8·2	2	16·4
8	8·8	4	35·2
9	9·1	2	18·2
10	9·4	4	37·6
11	9·5	1	9·5
		Total	226·3

$$\text{Area} = \tfrac{10}{3}(226\cdot3)$$

$$= 754\cdot3 \text{ sq. units}$$

Therefore work done = 754·3 kNm or kJ.

Now work done = change in kinetic energy.

$$\text{Therefore} \quad 754\cdot3 \times 10^3 = \tfrac{1}{2}m(v_2^2 - 0)$$

$$\text{therefore} \quad v_2^2 = \frac{754\cdot3 \times 10^3 \times 2}{100}$$

$$\text{therefore} \quad v_2 = 122\cdot8 \text{ m/s.}$$

STIFFNESS OF A SPRING

A spring is a good example of a force varying with distance. The strength of a spring is usually quoted as a spring stiffness and is measured in N/m. For example a spring of stiffness 2000 N/m requires a force of 2000 N to stretch (or compress) the spring one metre. Conversely if the spring is stretched (or compressed) by one metre there will be a force in the spring of 2000 N. The size of the force in the spring will be proportional to the size of the extension (or compression). If the increase in length doubles, the force in the spring will double; if the increase in length trebles, so the force in the spring will treble. The relationship holds up to the elastic limit on a pro-rata basis (it is most unlikely that a spring would actually extend by 1 m). In general

$$F = Sx \tag{1.14}$$

where S = spring stiffness in N/m

and x = change in length, which may be positive or negative

STRAIN ENERGY

The work done in stretching (or compressing) a spring is stored in the spring as strain energy. If the spring is released this energy will be used to return the spring to its original length. Thus

work done in stretching a spring = strain energy stored in spring

A graph of force F in the spring against extension x will result in a straight line, the area under which represents the strain energy stored in the spring (see diagram for example 1.10). Then

$$\text{work done} = \text{area under the graph}$$
$$= \tfrac{1}{2} Fx$$
$$= \tfrac{1}{2} Sx.x$$
$$= \tfrac{1}{2} Sx^2$$

Thus strain energy stored in spring = $\tfrac{1}{2} Sx^2$ joules.

Example 1.10

A spring balance has a stiffness of 800 N/m. Obtain the work done (W.D.) in stretching the spring for a load of 80 N, and the strain energy stored in the spring.

Since 800 N causes an extension of 1 m or 1000 mm, 80 N will cause an extension of 100 mm.

$$\text{W.D. in stretching spring} = \text{area beneath curve of work diagram}$$
$$= \tfrac{1}{2} \times 100 \times 80$$
$$= 4000 \text{ Nmm}$$
$$= 4 \text{ J}$$

i.e. strain energy stored in spring = W.D. in stretching spring = 4 J.

From the above example it will be seen that if the force varies *uniformly*, one can

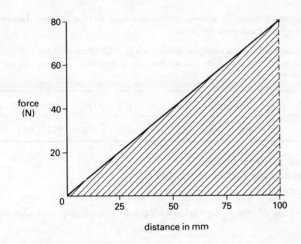

distance in mm

say: W.D. by a uniformly, variable force = average force × distance moved in direction of the force.

Problems 1.2

1 A car of mass 1 Mg travels at a speed of 50 km/h up an incline of 1 in 20, the frictional resistance being constant at 350 N. Find the amount of work done per minute:
  a against gravity
  b against friction.

2 a An aircraft, flying straight and level at a steady speed, covers 670 km in 1 hour 10 min. The resistance to motion is 9 kN. Calculate the work done per second against the resistance.
  b If the aircraft has a mass of 1 Mg and climbs at a rate of climb of 700 m/min, calculate the additional work done per second against gravity.

3 A mineshaft lift cage has a mass of 1500 kg and is supported by a cable of mass 3 kg/m. Determine the work done against gravity in raising the cage from 300 m to a depth of 150 m.

4 A train of mass 400 Mg travels up an incline of 1 in 200 at a constant speed of 50 km/h. The track resistance is 62 N/Mg. Find the total work done per minute.

5 The diameter of an engine piston is 300 mm and the stroke is 250 mm. During a working stroke the average pressure inside the cylinder is 69 bar. The engine does 400 working strokes per minute. How much work is done **a** per stroke and **b** per minute? (*Note*: 1 bar = 10^5 N/m^2)

6 In an experiment with a truck which was pulled over a horizontal table the tractive force F was measured by a spring balance at various distances x from the start. The readings were as follows:

F newtons	70	58	50	34	20	10
x metres	0	6	10	18	25	30

Plot the graph showing the relationship between the tractive force and the distance. Determine the work done in pulling the truck over the 30 m.

7 A spring balance was inserted between an engine and a truck in order to measure the pull exerted by the motor. The readings P Newtons of the balance were taken at distance S m from the starting point and were as follows.

P newtons	180	130	104	88	86	96	108	119·5	125	125	115
S metres	0	100	200	300	350	400	450	500	550	600	700

Plot a work diagram and from it obtain:
a work done in hauling the truck over the 700 m
b the mean pull exerted.

8 A force acting on a body of mass 2 kg varies with distance according to the formula.

$$F = 100 + 4x$$

Calculate the work done and the velocity after the body has moved a distance of 12 m from rest.

9 A uniform chain 50 m long has a mass of 9 kg/m and carries a load of mass 200 kg at its free end, the other end being attached to a winding drum.
a What work is done when the load is raised 50 m?
b If winding stops when only half the chain is wound up, how much work has been done?

1.6 The law of conservation of energy

If no energy enters or leaves a system, then the total energy within the system remains unchanged. Thus if a moving body experiences an increase in its kinetic energy, this will be at the expense of its potential energy, since

$$\text{total energy} = \text{kinetic energy (KE)} + \text{potential energy (PE)} = \text{constant}$$

For example, suppose a body of mass m kg falls freely from a height h_1 to the ground (Fig. 1.5). Total energy = KE + PE.

At A total energy $= 0 + mgh_1$

At B total energy $= \frac{1}{2}mv_2^2 + mgh_2$

At C total energy $= \frac{1}{2}mv_3^2$

Since the total energy is constant

$$mgh_1 = \frac{1}{2}mv_2^2 + mgh_2 = \frac{1}{2}mv_3^2.$$

On impact the total energy (KE + PE) of the body will be zero, all the energy being converted into heat, sound and strain energy.

Fig. 1.5

Example 1.11

A pendulum consists of a mass of 15 kg attached to the end of a pivoted lever 1·5 m long. The mass is held horizontally and released so that it swings down in a vertical circle.

a Assuming the pivot to be frictionless, calculate the maximum velocity attained by the mass.

b If the pivot is not frictionless and consumes 55 joules of energy between the two stationary positions of the mass, calculate the height to which the mass will rise on the opposite side to its starting point.

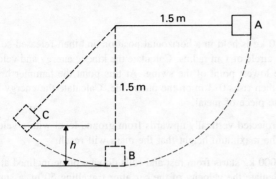

a At A, total energy = potential energy possessed by mass

$$= mgh$$
$$= 15 \times 9 \cdot 81 \times 1 \cdot 5$$
$$= 220 \cdot 7 \text{ joules}$$

At B, this energy is wholly converted into kinetic energy and the velocity is thus at its maximum value. Therefore

$$220 \cdot 7 = \tfrac{1}{2} \times 15 \times v^2$$

therefore $v = 5 \cdot 42$ m/s.

b At C, the mass is again stationary having swung to a maximum height h. Again the total energy is equal to the potential energy of the body.

However, of the original total of $220 \cdot 7$ joules, 55 joules has been given to overcome the friction of the pivot. Therefore remaining total energy = $165 \cdot 7$ joules. Thus

$$165 \cdot 7 = mgh$$

therefore $\quad h = \dfrac{165 \cdot 7}{15 \times 9 \cdot 81} = 1 \cdot 13$ m

Thus the mass will swing to a height of $1 \cdot 13$ m on the side opposite to its starting point.

Example 1.12

A truck of mass 1600 kg starts at rest and runs down an incline of 1/20. Calculate its velocity after it has travelled 30 m, if the total frictional resistances to motion are 250 N.

Potential energy lost = kinetic energy gained + work done against friction

$$\text{or} \quad mgh = \tfrac{1}{2}mv^2 + Fs$$

therefore $\quad 1600 \, g \, \tfrac{30}{20} = \tfrac{1}{2} \times 1600v^2 + 250 \times 30$

$$\text{therefore} \quad v^2 = \frac{2400g - 7500}{800}$$

$$\text{therefore} \quad v = 4 \cdot 48 \text{ m/s}$$

Problems 1.3

1   A hammer of mass 10 kg is held in a horizontal position and then released so that it swings in a vertical circle of 1 m radius. Calculate the kinetic energy and velocity of the hammer at the lowest point of the swing. At this point the hammer breaks a piece of metal and then rises $0 \cdot 3$ m on the other side. Calculate the energy consumed in breaking the piece of metal.

2   A mass of 12 kg is projected vertically upwards from ground level with a velocity of 8 m/s. Calculate the maximum height that the mass will reach.

3   A car of total mass 1000 kg starts from rest and rolls down a plane inclined at 30° to the horizontal. Calculate the velocity of the car after travelling 50 m, assuming no resistances to motion.

4   A car of mass 1000 kg starts from rest and accelerates up an incline of 1 in 10. After travelling 200 m its speed is 60 km/h. If the total frictional resistances to motion are 2500 N, calculate the work done by the car's engine and the tractive force if this is constant.

5   a   State the meaning of the terms strain energy and kinetic energy.
    b   A 2 kg body lies on a level table and is attached to one end of an elastic string. The other end of the string is tied to a fixture on the table so that when taut the string is horizontal. The coefficient of friction between the body and the table is $0 \cdot 5$ and it requires a force of 20 N to stretch the string $0 \cdot 6$ m. If the body is pulled away from the fixture until the tension in the string is 60 N and then released, calculate:
        i   the strain energy in the string
        ii the speed of the body when the string becomes slack.

6   A block of mass 400 kg is propelled up a slope of 1 in 20 by the release of a spring-loaded device which consists of two springs in compression. The strength of each spring is 15 kN/m and before release each spring is compressed $0 \cdot 6$ m. If the resistance of the block to sliding is 20 N calculate the speed of the block after it has moved 20 m up the slope.

7   a   State the principle known as the conservation of energy.
    b   A truck of mass 10 Mg at rest on an incline of 1 in 100 breaks loose and travels for 800 m down the incline. It continues on the level until it comes to rest. If the frictional resistance is 50 N/Mg determine:
        i   the speed of the truck at the bottom of the incline
        ii the distance travelled on the level

8   A truck moves along a level track and then up a gradient of 1 in 20. At a point 30 m up the gradient its velocity is $6 \cdot 7$ m/s up the gradient. The truck has a mass of 2 Mg and the resistance to motion is 180 N. Calculate:
    a   the distance the truck will continue up the gradient before coming to rest
    b   the distance along the level before the truck will stop, after running back down the gradient from the point where it came to rest.

9   A truck is moving down a track having a slope of 1 in 100 at a speed of 24 km/h
    when the wheels are locked by the application of the brakes. If the sliding friction
    between the wheels and the track is 1300 N per Mg,
    a how far will the truck move?
    b how long will it move, before coming to rest?

10 A train of mass 250 Mg, moving with a velocity of 65 km/h along a horizontal track,
    begins to climb up an incline of 1 in 80. During the climb the engine exerts a con-
    stant tractive force of 22 kN while the resistance to motion remains constant at 66 N
    per Mg. Determine:
    a how far the train will move?
    b how long it will move, before coming to rest?

11 **a** Give expressions for three forms of mechanical energy using the normal symbols
        and stating the units of each symbol used.
    b A body of mass 50 kg has a speed of 6 m/s at the bottom of a slope of 1 in 40
        and a speed of $1 \cdot 2$ m/s at the top. If the length of the slope is 6 m calculate:
        i the frictional resistance to sliding
        ii the time taken to come to rest on the level at the top of the slope if the resistance
            to sliding is 9 N.

12 **a** Define:
        i kinetic energy
        ii work.
    b A car of mass 900 kg reaches the foot of a slope of 1 in 20 with a speed of 20 m/s.
        Its engine is then shut off and it moves a distance of 30 m to the top of the slope.
        Immediately after this it runs a distance of 27 m down a slope of 1 in 18. If its
        final speed is 18 m/s find the resistance to motion, assumed constant.

13 A block of mass 50 kg is propelled up a slope of 1 in 20 by the release of a spring
    under compression. The strength of the spring is 16 kN/m and before release the
    spring is compressed 150 mm. If the resistance of the block to sliding is $4 \cdot 4$ N
    calculate:
    a the strain energy of the spring before release, and
    b the speed of the block after it has travelled 3 m up the slope.

14 A truck of mass 10 Mg moves from rest for 800 m down an incline of 1 in 220 against
    a frictional resistance of 35 N/Mg. At the bottom of the incline the truck strikes a
    pair of spring loaded buffers, on level ground, which are initially uncompressed. The
    spring stiffness is 36 N/mm. Find:
    a the potential energy of the truck at the top of the gradient
    b the kinetic energy of the truck when it strikes the buffers
    c the amount each buffer spring is compressed when the truck is brought to rest
    d the maximum force exerted on each spring.

15 The head of a hammer has a mass of 30 kg and is attached to one end of a handle
    1 m long and of negligible weight. The head and shaft of the hammer are held in
    a horizontal position initially. The heavy end is then released so that the head swings
    in a vertical circle of 1 m radius. At the lowest point of its swing the head strikes
    and breaks a metal specimen and moves on through an angle of 50° before coming
    to rest momentarily. Find:
    a the KE of the head of the hammer and its velocity immediately before striking
        the specimen
    b the energy used to break the specimen.

1.7   Power transmitted by a force

Power is the rate of doing work, i.e. the amount of work energy transferred per second. Thus

$$\text{work done} = Fs \text{ joules (from equation 5.6)}$$

$$\text{Rate of doing work} = \frac{d}{dt}(Fs) \text{ joules/second}$$

$$= F\frac{ds}{dt} \text{ assuming } F \text{ is constant.}$$

$$\text{but } \frac{ds}{dt} = \text{velocity}, v.$$

Therefore power = rate of doing work = Fv J/s or watts.

Example 1.13

A train travels at a steady speed of 60 km/h along a level track. If the total frictional resistances to motion are 22 kN calculate the power developed by the engine.

Since the train is travelling at a constant speed, the pull of the engine must equal the total frictional resistances to motion, i.e. pull of engine = 22 kN.

$$v = 60 \text{ km/h} = \frac{60 \times 10^3}{60 \times 60} = 16 \cdot 67 \text{ m/s}$$

$$\text{power} = Fv$$
$$= 22 \times 10^3 \times 16 \cdot 67$$
$$= 366\ 740 \text{ watts or } 366 \cdot 74 \text{ kW.}$$

Problems 1.4

1 The maximum speed of a car is 140 km/h. Calculate the power developed by the engine when travelling at top speed along a straight level road if the total resistances to motion are 3·5 kN.

2 A train of mass 3000 tonnes travels along a level track against frictional resistances which total 28 kN. Calculate:
   a the power developed by the engine when the train is travelling at a steady 80 km/h
   b the power developed at the instant that the speed is 20 km/h and the acceleration 1·2 m/s².

3 Calculate the power being developed by a tug which is manoeuvring a liner if the tension in the towing cable is 15 kN, the resistance to motion is 500 N and the time taken to pull the liner 120 metres in a straight line is 25 seconds.

1.8   Torque

Any force will have a turning effect about any point not on its own line of action. Thus, in Fig. 1.6 the force F has a turning point about the point O which is the product of the force and the perpendicular distance x of the line of action of the force from O. The turning effect is termed the **moment** of the force, i.e.

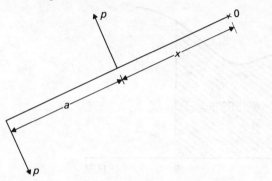

Fig. 1.6

moment $= Fx$ Nm

If two equal, parallel and exactly opposite forces act on a body with different lines of action then their effect is to produce a pure rotation. The moment of this **couple**, as the two forces are called, is independent of the distance from any fixed point. Thus in Fig. 1.7 taking moments about O.

Fig. 1.7

Net anti-clockwise moment $= p(a + x) - px$
$$= pa$$

Thus the moment of a couple, also referred to as a **torque**, is equal to the product of one of the forces and the perpendicular distance between the forces. In practice little reference is made to the force but only to the size of the torque, e.g. the torque applied to a body is T Nm.

1.9   Work done by a torque

Just as the work done by a force is the product of the force and the displacement, so then the work done by a torque is the product of the torque and the angular displacement i.e. work done by torque $= T\theta$ where θ is the angular displacement in radians.

Continuing the comparison with linear quantities, we learned earlier that the work done by a force results in an increase in the kinetic energy of the body, so the work done by a torque results in an increase in the angular kinetic energy.

Thus $T\theta = \frac{1}{2}I\omega_2^2 - \frac{1}{2}I\omega_1^2$ where I is the moment of inertia of the body, a term that depends not only on the magnitude of the mass but on its position in relation to the centre of rotation (see Appendix 4). ω is the angular velocity.

Example 1.14

The pedal of a bicycle is connected to a crank arm of length 150 mm. If the maximum force the rider is able to apply to the pedals is 70 newtons calculate the work done by the rider in making the crank turn through 360°.

Assuming two power strokes of 180° (or π radians) each are required (one on each pedal) then work done in one cycle is given by

$$
\begin{aligned}
\text{work done} &= \text{torque} \times \text{angular displacement} \\
&= 2 \times 70 \times 0\cdot15 \times \pi \\
&= 65\cdot98 \text{ joules.}
\end{aligned}
$$

1.10   Work done by a variable torque

Again, as in the linear case, the work done by a variable torque can be calculated from the area under the torque-angular displacement graph (Fig. 1.8):

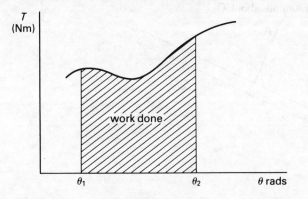

Fig. 1.8

Alternatively if the law connecting T and θ is known then the work done can be computed using

$$
\text{work done} = \int_{\theta_1}^{\theta_2} T \, d\theta
$$

Example 1.15

An elastic string fixed at one end is wound on a drum of diameter $0\cdot2$ m so that the tension in the string increases as the angle through which the drum turns increases. The torque on the drum then varies according to the law

$$
T = Sr^2\theta
$$

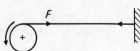

where S = stiffness of elastic string = 120 N/mm
and r = radius of drum.

Calculate the work done when the drum rotates through an angle of 120°.

$$
\theta = 120° = \frac{2\pi}{3} \text{ radians}
$$

$$
\begin{aligned}
\text{work done} &= \int_{\theta_1}^{\theta_2} T \, d\theta \\
&= \int_0^{2\pi/3} Sr^2\theta \, d\theta
\end{aligned}
$$

$$= \left[Sr^2 \frac{\theta^2}{2} \right]_0^{2\pi/3}$$

$$= \frac{Sr^2}{2} \left(\frac{2\pi}{3} \right)^2$$

$$= \frac{120 \times 10^3 \times (0 \cdot 1)^2 \times 4\pi^2}{2 \times 9}$$

$$= 2631 \cdot 9 \text{ joules} \quad \text{or} \quad 2 \cdot 63 \text{ kJ}$$

Example 1.16

The torque acting on an unbalanced flywheel varies with the angular displacement of the flywheel as shown in the table.

torque T Nm	0	200	400	600	800	1000	1200
angular displacement θ_{rad}	0	2·00	2·83	3·46	4·00	4·47	4·90

Calculate the work done by the torque when the flywheel rotates through 275°.

Plotting a graph of T against θ results in the curve shown. The work done is represented by the area under the curve. Using Simpson's rule with 6 strips, each of width 0·8 rads,

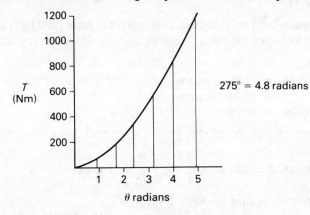

275° = 4.8 radians

(1) θ	(2) Ordinate	(3) Length	(4) Simpson's multiplier	Result
0	1	0	1	0
0·8	2	32	4	128
1·6	3	128	2	256
2·4	4	288	4	1152
3·2	5	512	2	1024
4·0	6	800	4	3200
4·8	7	1152	1	1152
			Total	6912

$$\text{Area} = \frac{h}{3}(A + 2B + 4C)$$

$$= \frac{0\cdot8}{3}(6912) = 1843\cdot2 \text{ sq. units}$$

Therefore work done by torque $= 1843\cdot2$ joules.

1.11 Power transmitted by a torque

Power, as stated previously, is the rate of work energy transfer. Thus, work done by torque $= T\theta$ joules.

$$\text{Rate of doing torsional work} = \frac{d}{dt}(T\theta)$$

$$= T\frac{d\theta}{dt}$$

$$= T\omega \text{ watts}$$

where ω is the angular velocity in rad/s.

Example 1.17

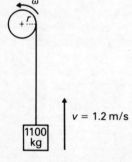

A winding drum of radius $0\cdot75$ m is required to raise a lift of mass 1100 kg at a steady velocity of $1\cdot2$ m/s. Calculate the power required to drive the drum, neglecting the mass of the cable.

$$
\begin{aligned}
\text{Tension in cable} &= 1100g \text{ newtons} \\
&= 10\ 791 \text{ newtons} \\
\text{torque on drum} &= 10\ 791 \times 0\cdot75 \text{ m} \\
&= 8093\cdot25 \text{ Nm}
\end{aligned}
$$

angular velocity of drum, $\omega = \dfrac{v}{r} = \dfrac{1\cdot2}{0\cdot75} = 1\cdot6$ rad/s

$$
\begin{aligned}
\text{Therefore power required} &= 8093\cdot25 \times 1\cdot6 \\
&= 12\ 949 \text{ watts} \quad \text{or} \quad 12\cdot95 \text{ kW}
\end{aligned}
$$

Note: In this example the power required could be found by considering the lift only. Thus, power required to raise lift at $1\cdot2$ m/s is

force on lift $\times$ velocity $= Fv = 1100g \times 1\cdot2 = 12\cdot95$ kW

Example 1.18

A chain of mass 6 kg per metre is wound on a drum of diameter $2\cdot5$ metres. A mass of 50 kg is attached to the free end and the system is released from rest with all the chain wrapped on the drum. Calculate the average power developed if the mass falls 12 metres in 9 seconds.

Tension in chain when x metres is wound off the drum = $(50 + 6x)g$ newtons.

Torque on drum = $(50 + 6x)g \times 1\cdot25$ Nm = $(50 + 6 \times 1\cdot25\theta)g \times 1\cdot25$ Nm since $x = r\theta$.

Therefore $T = 12\cdot26(50 + 7\cdot5\theta)$ Nm

Work done = $\displaystyle\int_{\theta_1}^{\theta_2} T\, d\theta$, and angle turned through by drum when 12 m of chain is

unwound = $\dfrac{12}{1\cdot25}$ rads = $9\cdot6$ rads

Therefore work done = $\displaystyle\int_0^{9\cdot6} 12\cdot26(50 + 7\cdot5\theta)\, d\theta$

$$= 12\cdot26\left[50\theta + 7\cdot5\,\frac{\theta^2}{2}\right]_0^{9\cdot6}$$

$$= 12\cdot26\left[50 \times 9\cdot6 + \frac{7\cdot5 \times 9\cdot6^2}{2}\right]$$

$$= 10\ 121\cdot9 \text{ joules}$$

Average power = work done per second = $\dfrac{10\ 121\cdot9}{9}$ = $1124\cdot66$ watts.

1.12 Power transmission using belts and spur gear systems

Often it is necessary to transmit the power developed in a shaft to another, parallel, shaft some distance away. For example, most overhead cam engines require some of the power developed at the crankshaft to be transmitted to the camshaft in order to operate the valve mechanism. Also, some power is required to drive a dynamo or alternator. This transfer of power is often done using belt, or chain and spur gear, systems (Fig. 1.9).

In the following analysis we will assume that no slip takes place between the belt and the pulleys and that the speeds of rotation are not high enough for centripetal effects to be significant. The analysis is applicable then to both belt and chain and spur gear systems. Consider the pulleys shown in Fig. 1.9.

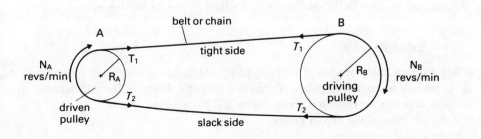

Fig. 1.9

Pulley A is the driven pulley, i.e. it rotates because of the action of the belt or chain on it. Pulley B is the driving pulley, it is *its* action which causes the belt or chain to move in the first place.

The driving pulley, rotating in a clockwise direction will cause a difference in tensions in the belt. In the case shown T_1 will be greater than T_2. This difference in tension is conveyed to the smaller pulley and so causes it to move. Let the radius and speed of rotation of the driving pulley be R_B and N_B respectively and the corresponding quantities for the smaller pulley be R_A and N_A.

Net torque on pulley A $= (T_1 - T_2)R_A$

Angular velocity of pulley A $= \omega_A = \dfrac{v}{R_A}$

where v is the circumferential velocity of the pulley or the **belt speed**.

Power required to drive pulley A $=$ torque $\times$ angular velocity

$$= (T_1 - T_2)R_A \times \dfrac{v}{R_A}$$
$$= (T_1 - T_2)v$$

In other words the power required to drive pulley A is the difference in belt tensions multiplied by the belt speed.

The student is left to determine for himself that the power supplied by pulley B also equals $(T_1 - T_2)\theta$.

Alternatively,

angular velocity of pulley A $= \omega_A = \dfrac{2\pi N_A}{60}$

therefore power required to drive pulley $=$ torque $\times$ angular velocity

$$= (T_1 - T_2)R_A \times \dfrac{2\pi N_A}{60}$$
$$= \dfrac{2\pi N_A R_A (T_1 - T_2)}{60} \text{ watts}$$

Using a similar reasoning for pulley B,

power developed by pulley B $= \dfrac{2\pi N_B R_B (T_1 - T_2)}{60}$ watts

Note: In high-speed applications or when the speed of the driving pulley changes rapidly, slip may occur, in which case the above analysis would be invalid.

Example 1.19

A belt drive assembly consists of a driving pulley of effective diameter 150 mm rotating at 80 rev/min and a driven pulley of effective diameter 320 mm. If the difference in tensions between the tight and slack side is 280 N calculate:

a the rev/min of the driven pulley

b the power transmitted

Considering the driving pulley,

$$80 \text{ rev/min} = 80 \times \frac{2\pi}{60} \text{ rad/s}$$

$$= 8 \cdot 378 \text{ rad/s}$$

$$\text{belt speed } v = r\omega = 0 \cdot 075 \times 8 \cdot 378$$

$$= 0 \cdot 628 \text{ m/s}$$

This speed is also the circumferential velocity of the driven pulley,

$$\text{therefore} \quad \omega = \frac{v}{r} = \frac{0 \cdot 628}{0 \cdot 160} = 3 \cdot 925 \text{ rad/s}$$

$$= \frac{3 \cdot 925 \times 60}{2\pi} \text{ rev/min}$$

$$= 37 \cdot 48 \text{ rev/min}$$

$$\text{power transmitted} = (T_1 - T_2)v$$

$$= 280 \times 0 \cdot 628$$

$$= 175 \cdot 84 \text{ watts}$$

1.13   Power to overcome friction

In section 1.5 we discussed the work done to overcome friction. When power is being transmitted, say via a shaft, there is work being done to overcome the friction in bearings holding the shaft. In other words there will be a power loss in the bearings. In fact whenever friction is present, and in real situations friction is *always* present, power is required to overcome its effects.

Example 1.20

Calculate the frictional torque resisting the rotation of a $0 \cdot 025$ m diameter shaft if the load on a journal bearing retaining the shaft is 500 N and the coefficient of friction between the shaft and the bearing is $0 \cdot 3$.

If the shaft rotates at 1500 rev/min calculate the power absorbed by friction.

$$\text{frictional resistance of bearing} = \mu N, \text{ where N is the load on the bearing}$$

$$= 0 \cdot 3 \times 500$$

$$= 150 \text{ N}$$

$$\text{frictional torque} = \text{frictional resistance} \times \text{radius}$$

$$= 150 \times 0 \cdot 0125$$

$$= 1 \cdot 875 \text{ Nm}$$

$$\text{power absorbed} = \text{frictional torque} \times \text{angular velocity}$$

$$= T\omega$$

$$= 1 \cdot 875 \times \frac{2\pi \times 1500}{60}$$

$$= 294 \cdot 5 \text{ watts}$$

Problems 1.5

1   A wheel is caused to rotate by a torque of 200 Nm applied to the rim. Calculate the work done by the torque in moving the wheel a distance equal to its circumference.

2   An engine running at 500 rev/min carries a load of 200 N at the rim of the flywheel which has a radius of $0 \cdot 4$ m. Calculate the power developed by the engine.

3   Calculate the number of revolutions per second made by the winding drum in bringing a pit cage of total mass 1500 kg to the surface. The power required by the winding gear is 200 kW and the diameter of the drum is 6 m.

4   The torque applied by a spanner to a tight nut varies according to the law

   $T = 1 \cdot 2(\theta + 2)$, where θ is the angle turned through in radians.

   Calculate the work done in tightening the nut so that it rotates a further $100°$.

5   The torque required to move a flywheel is measured at various angular displacements and the following results are noted:

torque (Nm)	220	226	232	238	244	250	256	262
angular displacement (rads)	0	0·5	1·0	1·5	2·0	2·5	3·0	3·5

   Determine the work done to move the flywheel through half a revolution.

6   The torque acting on a winding drum varies with the angular displacement of the drum as shown in the following table:

torque (Nm)	100	150	220	295	420	800
angular displacement (rads)	0	$\pi/3$	$4\pi/3$	2π	$8\pi/3$	4π

   Determine the work done by the torque when the drum rotates through $720°$.

7   A chain of mass $5 \cdot 2$ kg/m is wound on a drum of diameter $0 \cdot 65$ m. The free end of the chain is attached to a mass of 750 kg on the ground 10 m below the level of the drum. Deduce an expression for the torque required on the drum to lift the mass in terms of θ, the angular displacement of the drum, and the work done by this torque in raising the mass 8 m.

8   The diagram shows the torque on the crankshaft of an engine as a function of the angular displacement of the shaft. Find
   a the work done per revolution of the shaft
   b the average torque
   If the engine speed is 300 rev/min find the average power transmitted.

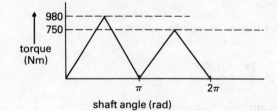

shaft angle (rad)

9 Calculate the power required to drive a conveyor belt at 250 mm/sec if the tension on the slack side of the belt is 20 N and on the tight side 250 N.

10 An engine developing 4 kW is used to drive a pulley of effective diameter 25 cm at 300 rev/min. A belt connects this pulley to another of effective diameter 40 cm. Calculate the difference in belt tension across a pulley, and the torque applied to the second pulley.

11 The load from a shaft collar on a thrust bearing of 0·1 m mean diameter is 20 kN and the coefficient of friction between the shaft and the bearing is 0·40. Calculate
a the frictional torque resisting the rotation
b the power absorbed by the bearing when the shaft is rotating at 150 rev/min.

12 A dry journal bearing carries a load of 5 kN from a shaft of diameter 0·08 m. If the coefficient of friction between the shaft and the bearing is 0·65 and the shaft rotates at 180 rev/min, calculate the power absorbed by the bearing. If the journal is lubricated to reduce the coefficient of friction to 0·28 calculate the new power absorbed.

2 Deformation of materials

The contents of this chapter enable you to use stress, strain and elasticity in problems.

2.1 Simple direct stress and strain

The simplest stress situation occurs when a direct axial load is applied to a bar of material of constant cross-section (Fig. 2.1).

The stress, σ, is defined as:

$$\sigma = \frac{\text{load}}{\text{area}} = \frac{F}{A} \text{ N/m}^2 \qquad (2.1)$$

and is positive if it is tensile and negative if it is compressive.

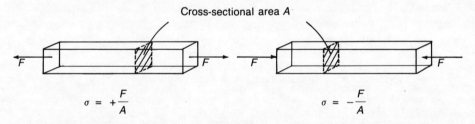

Cross-sectional area A

$$\sigma = +\frac{F}{A} \qquad\qquad \sigma = -\frac{F}{A}$$

Fig. 2.1

The applied stress will always produce a deformation in the form of a change of length of the bar. The change of length is usually described as a fraction of the original length, so that the deformation is stated as a strain, ϵ, where:

$$\epsilon = \frac{\text{change of length}}{\text{original length}} = \frac{\delta l}{l} \qquad (2.2)$$

Hooke's Law states that for perfect materials the stress is proportional to the strain up to the elastic limit, so that

$$\sigma \propto \epsilon$$
$$\text{or} \quad \sigma = E\epsilon \qquad (2.3)$$

where E is a constant for the material known as Young's Modulus of Elasticity.

Since ϵ has no units, being simply the ratio of two lengths, the E has the same units as σ, i.e. N/m^2.

Example 2.1

Calculate the change in length of a mild steel bar, 2 m long and 50 mm diameter, when subjected to a tensile load of 40 kN.

Take E for mild steel to be 208 GN/m^2.

$$\text{Since } E = \frac{\sigma}{\epsilon} = \frac{F/A}{\delta l/l} = \frac{Fl}{A\delta l}$$

then, re-arranging

$$\delta l = \frac{Fl}{EA}$$

$$= \frac{40 \times 10^3 \times 2}{208 \times 10^9 \times \pi \times \dfrac{(0.05)^2}{4}}$$

$$= 0 \cdot 000196 \text{ m}$$
$$= 0 \cdot 196 \text{ mm}$$

2.2   Stress in bars with change of section

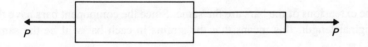

Fig. 2.2

Consider the bar shown in Fig. 2.2, subject to a tensile load P. The bar is turned down to a smaller cross-sectional area over part of its total length. The following points are true.

a The load P is the same for each part of the bar.

b Since stress $= \dfrac{\text{load}}{\text{area}}$, the stress is greater in the narrower portion of the bar.

c Since $\dfrac{\text{stress}}{\text{strain}} = E$, Young's Modulus, then, strain $= \dfrac{\text{stress}}{E}$, and thus the strain is greater in the narrower portion of the bar.

2.3   Composite bars

A composite or compound bar consists of two or more parallel bars of different materials rigidly connected together. When such a composite bar is subjected to a pure axial load in one direction only, the bar deforms by a different amount (and possibly in a different manner) than either of the component bars would have done if they had been subjected to the same load individually. For the composite bar two important facts are always true:

a The load applied to the composite bar is *shared* by each component bar.

b Since the component bars are rigidly connected they all experience the *same* extension.

For example, consider the composite bar consisting of three component bars (Fig. 2.3). The outer bars are both made from material A and the inner bar from material B. The symmetrical arrangement means that the applied tensile force P will have a pure tension effect only without a tendency to bend the bar.

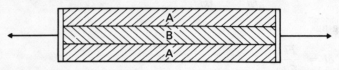

Fig. 2.3

Applying the basic facts to the bar:

1 The load is shared by each bar, i.e.

$$F_A + F_B = P \tag{2.4}$$

where F_A is the total force in both bars of material A

or $\quad \sigma_A A_A + \sigma_B A_B = P \tag{2.5}$

where σ_A and σ_B are the stresses in the bar A and bar B respectively, and A_A is the total area of cross-section of the two bars of material A. A_B is the area of cross-section of bar B.

2 The extensions of the bars are the same. Since the component bars have the same original length, this means that the strains in each bar will be the same, i.e.

$$\epsilon_A = \epsilon_B \tag{2.6}$$

or $\quad \dfrac{\sigma_A}{E_A} = \dfrac{\sigma_B}{E_B} \tag{2.7}$

where E_A and E_B are the values of Young's modulus of elasticity for the two materials.

Example 2.2

A bar of length $3 \cdot 0$ m has a diameter of 50 mm over half its length and a diameter of 25 mm over the other half. If $E = 206$ GN/m^2 and the bar is subjected to a pull of 50 kN, find the stress in each section and the total extension of the bar.

Each section of the bar is subjected to the same load $= 50$ kN.

Therefore stress in large cross-section $= \dfrac{F}{A} = \dfrac{50 \times 10^3}{\pi (0 \cdot 05)^2 / 4} = 25 \cdot 46$ MN/m^2

stress in smaller cross-section $= \dfrac{50 \times 10^3}{\dfrac{\pi (0 \cdot 025)^2 l}{4}} = 101 \cdot 86$ MN/m^2

strain in large cross-section $= \dfrac{\sigma}{E} = \dfrac{25 \cdot 46 \times 10^6}{206 \times 10^9} = 0 \cdot 1236 \times 10^{-3}$

Therefore change in length of larger section

$$= 0.1236 \times 10^{-3} \times 1.5 = 0.1854 \times 10^{-3} \text{ m}$$

strain in smaller cross-section

$$= \frac{\sigma}{E} = \frac{101.86 \times 10^6}{206 \times 10^9} = 0.4945 \times 10^{-3} \text{m}$$

Therefore change in length of smaller section

$$= 0.4945 \times 10^{-3} \times 1.5 = 0.7417 \times 10^{-3} \text{ m}$$

Total change in length of bar $= (0.1854 + 0.7417)10^{-3}$ m $= 0.927$ mm

Example 2.3

A mild steel bar 50 mm in diameter and 10 cm long is fitted inside a copper tube of internal diameter 50 mm, external diameter 60 mm and length 10 cm. The bar and the tube are both rigidly connected to plates at each end and the whole is subjected to a tensile load of 60 kN. Calculate the stresses in each material and the extension of the composite bar.

$$(E_{\text{steel}} = 200 \text{ GN/m}^2, E_{\text{copper}} = 110 \text{ GN/m}^2)$$

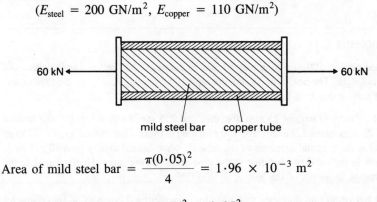

60 kN ← → 60 kN

mild steel bar   copper tube

Area of mild steel bar $= \dfrac{\pi(0.05)^2}{4} = 1.96 \times 10^{-3} \text{ m}^2$

Area of copper tube $= \dfrac{\pi(0.06^2 - 0.05^2)}{4} = 8.64 \times 10^{-4} \text{ m}^2$

The load is shared by the steel and the copper,

$$\text{therefore} \quad \sigma_s A_s + \sigma_c A_c = P \tag{1}$$

The strain in the steel equals the strain in the copper,

$$\text{therefore} \quad \frac{\sigma_s}{E_s} = \frac{\sigma_c}{E_c} \quad \text{or} \quad \sigma_s = \frac{E_s}{E_c} \times \sigma_c \tag{2}$$

Substituting (2) in (1) $\dfrac{E_s}{E_c} \times \sigma_c A_s + \sigma_c A_c = P$

$$\sigma_c\left(\frac{E_s}{E_c} \times A_s + A_c\right) = P$$

$$\sigma_c = \cfrac{P}{\cfrac{E_s}{E_c} \times A_s + A_c}$$

$$= \cfrac{60 \times 10^3}{\cfrac{200 \times 10^9}{110 \times 10^9} \times 1\cdot96 \times 10^{-3} + 8\cdot64 \times 10^{-4}}$$

$$= 13\cdot55 \text{ MN/m}^2$$

Substituting in (2), $\sigma_s = 24\cdot64$ MN/m^2.

The strain in the bar $= \dfrac{\sigma_s}{E_s} = \dfrac{\sigma_c}{E_c} = \dfrac{13\cdot55 \times 10^6}{110 \times 10^9} = 1\cdot23 \times 10^{-4}$

Now, extension = strain $\times$ original length

$$= 1\cdot23 \times 10^{-4} \times 0\cdot1 \times 10^3 = 0\cdot0123 \text{ mm}$$

Problems 2.1

1 A duralumin tie, 500 mm long and 40 mm in diameter, has an axial hole drilled out along its length. The hole is of 25 mm dia. and 100 mm long. Calculate the total extension of the tie due to a load of 180 kN. $E = 82\cdot5$ GN/m^2.

2 A solid cylindrical bar, of 25 mm diameter and 225 mm long, is welded to a hollow tube of 25 mm internal diameter and 150 mm long to make a bar of total length 375 mm. Determine the external diameter of the tube if, when loaded axially by a 40 kN load, the stress in the solid bar and that in the tube are to be the same. Hence calculate the total change in length of the bar. $E = 206$ GN/m^2.

3 A steel bar of 50 mm diameter and 300 mm long is turned down to 38 mm diameter for a length of 75 mm and reduced to 25 mm diameter for a further length of 100 mm. If E for the material $= 206$ GN/m^2, calculate when carrying a load of 70 kN:
   a the stress in each portion of the bar
   b the total extension of the bar

4 A rectangular timber tie, 175 mm by 75 mm, is reinforced by two aluminium bars 60 mm wide by 6 mm thick, firmly fixed to each side. Calculate the stresses in the timber and reinforcement when the tie carries an axial load of 300 kN. E for timber $= 15$ GN/m^2; E for aluminium $= 90$ GN/m^2.

5 A concrete column having modules of elasticity 20·6 GN/m^2 is reinforced by 2 steel bars of 25 mm diameter having a modulus of 206 GN/m^2. Calculate the dimensions of a square section column if the stress in the concrete is not to exceed 6·8 N/mm^2 and the load is to be 400 kN.

6 A cylindrical mild steel bar of 38 mm diameter and 150 mm long is enclosed by a bronze tube of the same length having an outside diameter of 65 mm and inside diameter of 38 mm. This compound strut is subjected to an axial compressive load of 200 kN.

Find:
  a the stress in the steel rod
  b the stress in the bronze tube
  c the shortening of the strut
For steel $E = 206$ GN/m^2. For bronze $E = 96$ GN/m^2.

7 A compound assembly is formed by brazing a brass sleeve on to a solid steel bar of 50 mm diameter. The assembly is to carry a tensile axial load of 250 kN. Find the cross-sectional area of the brass sleeve so that the sleeve carries 30% of the load. Find also for the composite bar the stresses in the brass and steel.
E for brass $= 82 \cdot 5$ GN/m^2. E for steel $= 206$ GN/m^2.

8 A steel tube 150 mm long, 75 mm external diameter and 50 mm internal diameter is bushed with a bronze tube of the same length, 50 mm external diameter and 25 mm internal diameter. If the compound tube is subjected to an axial compressive load of 50 kN, find:
  a the stress in the steel tube
  b the stress in the bronze bush
  c by how much the tube will be shortened
E for steel $= 206$ GN/m^2, E for bronze $= 95$ GN/m^2.

2.4   Composite bars subjected to uniform temperature change

When any solid is subjected to a temperature increase an expansion takes place. Any linear dimension l of the solid will increase in length to a new length $l + x$, where x, the extension will depend on the value of the temperature change, the material of which the solid is made, and the original length of the solid.

$$\text{Thus } x = \alpha l t \qquad (2.8)$$

where α is the coefficient of linear expansion, which is constant for a particular material.

$$\text{Thus the new length } = l + \alpha l t. \qquad (2.9)$$

When a composite bar of two or more different materials, each of different coefficients of linear expansion, is heated uniformly, the component bars will attempt to expand by different amounts. They will be prevented from doing so because they are rigidly connected together.

Consider the example of the composite bar used previously, subjected to a uniform temperature increase. If the component bars were not rigidly connected together, each would expand freely by different amounts, depending on their particular coefficient of expansion.

Thus the bars of material A would expand by an amount $x_A = \alpha_A l t$, and the bar of material B by an amount $x_B = \alpha_B l t$.

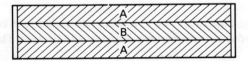

Fig. 2.4

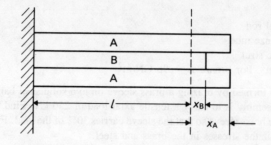

Fig. 2.5

If α_A is greater than α_B then the unconnected bars would look like Fig. 2.5 (greatly exaggerated). Since in fact the bars *are* rigidly connected, the bars of material A will expand not by an amount x_A but by a lesser amount x, being held back by the bar of material B. Similarly the bar of material B will expand not by an amount x_B but by a greater amount x, being extended further by the effects of material A. The net result will look like Fig. 2.6a.

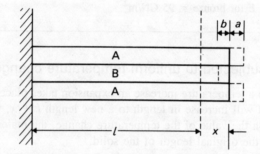

Fig. 2.6a

In effect, both the bars of material A have been *compressed* each by an amount a and are thus subject to a compressive stress σ_A. Bar B on the other hand has been *extended* by an amount b and is thus subject to a tensile stress σ_B (Fig. 2.6b).

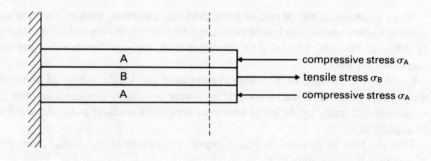

Fig. 2.6b

Examination of the diagrams in Figs 2.5 and 2.6a reveals the relationship

$$x_A - x_B = a + b$$

That is, the difference in the *free* expansions due to a temperature change is equal to the sum of the changes in length due to the compressive and tensile stresses. This may also be written

$$\alpha_A lt - \alpha_B lt = a + b$$

$$\text{or} \quad (\alpha_A - \alpha_B)t = \frac{a}{l} + \frac{b}{l}$$

$$= \epsilon_A + \epsilon_B$$

$$= \frac{\sigma_A}{E_A} + \frac{\sigma_B}{E_B} \tag{2.10}$$

where ϵ_A and ϵ_B are the strains in materials A and B respectively.

The composite bar is internally in equilibrium, the tensile *force* that A exerts on B being equal but opposite to the compressive *force* that B exerts on A. Thus

$$\sigma_A A_A = \sigma_B A_B \tag{2.11}$$

Equations (2.10) and (2.11) are usually sufficient for finding the stresses in the bars.

A compound bar subjected to both an applied load and a rise in temperature will experience a stress due to the applied load together with a temperature stress. The analysis for each part can be made separately and the results superimposed, i.e. the total strain in the bar will be the sum of the strain due to loading and temperature change.

Example 2.4

A cylinder of steel fits inside a cylinder of copper. The two cylinders are of equal length of 45 mm and are rigidly connected together at their ends. The copper cylinder has an internal diameter of 12 mm and an external diameter of 15 mm, and the steel cylinder an internal diameter of 9 mm and an external diameter of 12 mm. Calculate the stresses set up in each cylinder due to a temperature rise of 20°C.

$$(E_{steel} = 200 \text{ GN/m}^2, \ E_{copper} = 110 \text{ GN/m}^2,$$
$$\alpha_{steel} = 11 \times 10^{-6}/°C, \ \alpha_{copper} = 18 \times 10^{-6}/°C)$$

The *free* expansion of the copper $= \alpha_c lt$

$$= 18 \times 10^{-6} \times 0 \cdot 045 \times 20 = 16 \cdot 2 \times 10^{-6} \text{ m}$$

The *free* expansion of the steel $\alpha_s lt$

$$= 11 \times 10^{-6} \times 0 \cdot 045 \times 20 = 9 \cdot 9 \times 10^{-6} \text{ m.}$$

Difference in free expansions $= (\alpha_c - \alpha_s)lt = 6 \cdot 3 \times 10^{-6} \text{ m.}$

This difference divided by l is equal to the sum of the strains in the steel and copper, i.e.

$$\frac{6 \cdot 3 \times 10^{-6}}{0 \cdot 045} = \frac{\sigma_s}{E_s} + \frac{\sigma_c}{E_c}$$

$$\text{or} \quad 140 \times 10^{-6} = \frac{\sigma_s}{200 \times 10^9} + \frac{\sigma_c}{110 \times 10^9} \tag{1}$$

Also the tensile load in the steel = compressive load in copper

$$\sigma_s A_s = \sigma_c A_c$$

$$\text{or} \quad \sigma_s = \frac{A_c}{A_s} \times \sigma_c = \frac{\dfrac{\pi(0 \cdot 015^2 - 0 \cdot 012^2)}{4}}{\dfrac{\pi(0 \cdot 012^2 - 0 \cdot 009^2)}{4}} \times \sigma_c$$

therefore $\sigma_s = 1 \cdot 286 \sigma_c$ \hfill (2)

Substituting (2) in (1)

$$140 \times 10^{-6} = \frac{\sigma_c}{10^9} \left(\frac{1 \cdot 286}{200} + \frac{1}{110} \right)$$

$$140 \times 10^{-6} = \frac{\sigma_c}{10^9} (0 \cdot 015\ 52)$$

therefore $\sigma_c = 9 \cdot 02 \text{ MN/m}^2$

Substituting this figure in (2) gives

$$\sigma_s = 11 \cdot 6 \text{ MN/m}^2.$$

Problems 2.2

Take $E_{steel} = 200 \text{ GN/m}^2$, $E_{copper} = 110 \text{ GN/m}^2$, $E_{bronze} = 120 \text{ GN/m}^2$.
$\alpha_{steel} = 11 \times 10^{-6}/°C$, $\alpha_{copper} = 18 \times 10^{-6}/°C$, $\alpha_{bronze} = 20 \times 10^{-6}/°C$.

1 A steel tube of 25 mm outside diameter and 3 mm thick has a bronze rod of 15 mm diameter inside it and rigidly joined to it at each end. If at 15°C there is no stress, calculate the stress in the rod and the tube when their temperature is raised to 200°C.

2 A steel tube of 25 mm external diameter and 15 mm internal diameter encloses a copper rod of 12 mm diameter to which it is rigidly fixed. Calculate the stress in the tube and the rod when their temperature is raised to 200°C if there was no stress when their temperature was 18°C.

3 A copper tube 18 mm external diameter, 12 mm internal diameter and 4 m long has its temperature raised from 15° to 60°C, and the ends are then secured to prevent contraction. If the bar then cools to its original temperature calculate the pull on the bar.

4 Steel railway lines are welded together at 16°C. Calculate the stresses that will be produced in the rails when heated by the sun to 30°C.

5 A steel steam pipe is fitted in two brackets 1·0 m apart so as to allow free longitudinal expansion and contraction as the temperature changes. Due to corrosion the brackets become rigidly fixed to the pipe, so that during a 100°C temperature rise the free expansion of the pipe is limited to a half of its normal value. Calculate the resulting longitudinal stress in the pipe.

2.5   Shear

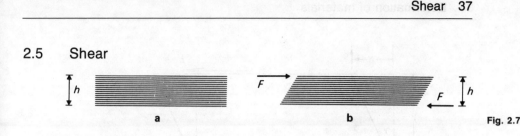

a   b   Fig. 2.7

A material is said to be in shear if the forces applied to it tend to slide one face of the material over an adjacent face. To illustrate exactly what is happening, consider a pack of playing cards to represent a block of material (Fig. 2.7a). A couple Fh applied parallel to the surface of the pack would cause the pack to deform as shown (Fig. 2.7b), that is, providing each card moves by the same amount relative to the next. This is exactly what happens to a block of solid material to which a shearing force is applied.

Shear forces are extremely common in engineering. For example, the rivets holding two flat pieces of metal close together are in shear if a tensile force is applied to the plates (Fig. 2.8a). Sufficient shearing force could cause the rivet to fail (Fig. 2.8b).

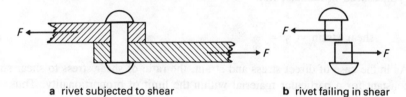

a rivet subjected to shear   b rivet failing in shear   Fig. 2.8

A shaft subjected to a pure torque is subjected to a shearing force. Again, too great a value of torque could cause the shaft to fail in shear, for example as a bolt might do if overtightened with a spanner.

A simply supported beam subjected to vertical loads is caused to sag and as a result is subjected to shear forces in two directions at right angles (Fig. 2.9).

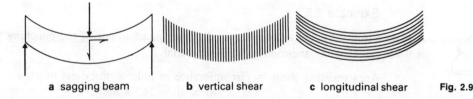

a sagging beam   b vertical shear   c longitudinal shear   Fig. 2.9

The **shear stress**, τ, which arises as a result of an applied shearing load, F, is defined as follows:

$$\tau = \frac{F}{A} \tag{2.12}$$

where A is the area resisting shear.

This area is the area over which the shearing force acts, i.e. the area *parallel* to the direction of the shear force. It would be the area of one surface of one card in the playing card pack.

A shear stress will produce a deformation in a material as in Fig. 2.10. If x is the amount of relative shift between two layers subjected to shearing force F and a distance

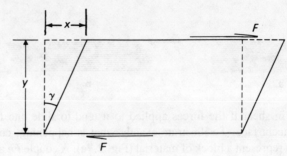

Fig. 2.10

y apart, then

$$\text{shear strain} = \frac{x}{y} = \tan \gamma$$

Since x will be very small compared with y, then $\tan \gamma$ will be approximately equal to γ measured in radians, i.e.

$$\text{shear strain} = \gamma = \frac{x}{y} \tag{2.13}$$

As in the case of direct stress and strain, the ratio of shear stress to shear strain is a constant for a particular material within the limit of proportionality. Thus

$$\frac{\text{shear stress}}{\text{shear strain}} = \text{constant, } G$$

The constant G, is called the **modulus of rigidity** or **shear modulus**, and since strain has no units it has the same units as shear stress, i.e. N/m^2.

Example 2.5

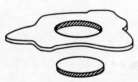

A plate 6 mm thick has a hole 40 mm diameter punched out. If the punching force is 9 kN calculate the shear stress in the material.

Area resisting shear = circumference of hole × thickness of plate
$$= \pi \times 40 \times 6 = 240 \, \pi \text{ mm}^2.$$

$$\text{shear stress} = \frac{\text{shear force}}{\text{area}} = \frac{9 \times 10^3}{240\pi} = 11 \cdot 94 \text{ N/mm}^2$$

Example 2.6

Calculate the power which can be safely transmitted at 500 rev/min via a flanged coupling fastened with 6 bolts of 12 mm diameter on a pitch circle diameter of 250 mm. The maximum shear stress of the material is 385 N/mm² and a factor of safety of 5 is to be used. (The torque due to friction between the coupling faces is to be ignored.)

Shearing force in all bolts = shear stress
 × cross-sectional area of bolt
 × number of bolts

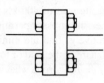 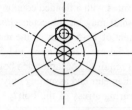

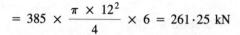

$$= 385 \times \frac{\pi \times 12^2}{4} \times 6 = 261 \cdot 25 \text{ kN}$$

Including safety factor max. shearing force is $\frac{1}{5}$ of this $= 52 \cdot 25$ kN

Torque transmitted by bolts = max shearing force
$$\times \text{ radius of pitch circle}$$
$$= 52 \cdot 25 \times 0 \cdot 125 = 6 \cdot 53 \text{ kNm.}$$

Maximum power transmitted $= T\omega = T \times \dfrac{2\pi N}{60} = \dfrac{6 \cdot 53 \times 10^3 \times 2\pi \times 500}{60}$

$$= 342 \text{ kW}$$

Problems 2.3

1 Calculate the maximum thickness of plate which can be sheared on a guillotine if the maximum shearing strength of the plate is 250 N/mm^2 and the maximum force the guillotine can exert is 200 kN. The width of the plate is 1 m.

2 A rectangular hole 50 mm by 62 mm is punched in a steel plate 6 mm thick. The maximum shearing stress of the plate is 200 N/mm^2. Calculate the load on the punch.

3 Calculate the maximum diameter of hole which can be punched in 1·5 mm plate if the punching force is limited to 40 kN. The plate is aluminium having a max. shear stress of 93 N/mm^2.

4 A bar is cut at 30° to its axis and joined by two 12 mm dia. bolts as shown. If the axial pull in the bar is 80 kN, calculate the tensile and shear stresses in each bolt.

5 A solid coupling is to transmit 225 kW at 600 rev/min. The coupling is fastened with 6 bolts on a pitch circle diameter of 200 mm. If the maximum shear stress of the bolt material is 310 N/mm^2, calculate the bolt diameter required. The factor of safety is to be 4.

6 A shaft is to transmit 185 kW at 600 rev/min through solid coupling flanges. There are four coupling bolts, each of 12 mm diameter. If the shear stress in each bolt is to be limited to 39 N/mm^2, calculate the minimum diameter of the circle at which the bolts are to be placed.

7 A gear wheel 25 mm thick is shrunk on to a 50 mm diameter shaft so that the radial pressure at the circle of contact is 6·8 N/mm^2. The coefficient of friction between gear wheel and shaft is 0·2 and they are also prevented from relative rotation by a key 6 mm wide by 25 mm long. If the shaft transmits 15 kW at 100 rev/min calculate the shear stress in the key.

8   A shaft is to be fitted with a flanged coupling having 8 bolts on a pitch circle of diameter 150 mm. The shaft may be subjected *either* to a direct tensile load of 400 kN *or* to a twisting moment of $17 \cdot 5$ kNm. If the max tensile and shearing stresses permissible in the bolt material are 124 N/mm^2 and 54 N/mm^2 respectively, find the minimum diameter of bolt required. Assume each bolt takes an equal share of the load or torque. Using this bolt diameter and assuming only one bolt to carry the full torque what would then be the shearing stress of the bolt?

2.6   Poisson's ratio

A bar of material subjected to a tensile stress will get longer and thinner (Fig. 2.11a). A bar subjected to a compressive stress will get shorter and fatter (Fig. 2.11b).

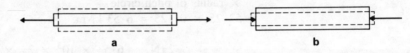

Fig. 2.11

a   b

In general a material subjected to an applied direct stress will not only suffer a change in dimension in the direction of the stress, but also in a direction at right angles to the applied stress. For example a bar of material of length x subjected to a tensile stress σ_x will experience an extension δx where

$$\frac{\delta x}{x} = \frac{\sigma_x}{E}$$

At the same time, the dimensions at right angles (i.e. the y and z directions) are altered so that the strain in these directions is some proportion of the strain in the x-direction.

$$\frac{\delta y}{y} = \nu \frac{\sigma_x}{E}$$

and $$\frac{\delta z}{z} = \nu \frac{\sigma_x}{E}$$

The constant ν is called **Poisson's ratio**, and is defined as follows:

$$\nu = \frac{\text{strain in direction at right angles to applied stress}}{\text{strain in direction of applied stress}}$$

hence $$\nu = \frac{\delta y / y}{\sigma_x / E} = \frac{\epsilon_y}{\epsilon_x} \quad \text{or} \quad \nu = \frac{\delta z / z}{\sigma_x / E} = \frac{\epsilon_z}{\epsilon_x}$$

Example 2.7

A mild steel bar 2 m long and 50 mm diameter is subjected to a direct tensile load of 250 kN. Calculate the change in diameter of the bar, if $E = 200$ GN/m^2 and $\nu = 0 \cdot 3$

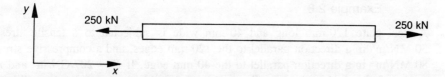

The strain in the x-direction, $\epsilon_x = \dfrac{\sigma_x}{E}$

$$= \frac{F}{EA}$$

$$= \frac{250 \times 10^3}{200 \times 10^9 \times \pi \left(\dfrac{0 \cdot 05}{4}\right)^2}$$

$$= 636 \cdot 6 \times 10^{-6}$$
$$= 636 \cdot 6 \ \mu \ \text{strain}$$

Using Poisson's ratio

$$\epsilon_y = \nu \epsilon_x$$
$$= -0 \cdot 3 \times 636 \cdot 6 \times 10^{-6}$$
$$= \underline{-191 \cdot 0 \times 10^{-6}}$$

Now $\epsilon_y = \dfrac{\text{change in diameter}}{\text{original diameter}}$

$\therefore$ change in diameter $= \epsilon_y \times$ diameter
$$= -191 \cdot 0 \times 10^{-6} \times 50$$
$$= \underline{9 \cdot 55 \times 10^{-3} \ \text{mm}}$$

If, in addition to the stress σ_x in the x-direction, there is also an applied stress σ_y in the y-direction then there will be a strain σ_y/E in the y-direction plus strains of $\nu \sigma_y/E$ in the x and z directions.

Thus the *net* strains in the x, y and z directions will be:

$$\epsilon_x \qquad = \frac{\sigma_x}{E} \qquad\qquad - \nu \frac{\sigma_y}{E}$$

(Total strain = (strain in x-direction   − (strain in x-direction
in x-direction)   due to σ_x)   due to σ_y)

$$\epsilon_y \qquad = \frac{\sigma_y}{E} \qquad\qquad - \nu \frac{\sigma_x}{E}$$

(Total strain = (strain in y-direction   − (strain in y-direction
in y-direction)   due to σ_y due to σ_x)

$$\epsilon_x \qquad = -\nu \frac{\sigma_x}{E} \qquad\quad - \nu \frac{\sigma_y}{E}$$

(Total strain = − (strain in z-direction   − (strain in z-direction
in z-direction)   due to σ_x)   due to σ_y)

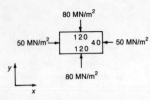

Example 2.8

A steel plate 120 mm long and 40 mm wide is subjected to a tensile stress of 50 MN/m^2 in a direction parallel to the 120 mm edges, and a compressive stress of 80 MN/m^2 in a direction parallel to the 40 mm edge. If $E = 207$ GN/m^2 and ν for steel $= 0 \cdot 3$ calculate the new dimensions of the plate in the stressed condition.

Let $\sigma_x = 50$ MN/m^2 and $\sigma_y = -80$ MN/m^2
(Note the sign convention for tensile and compressive stresses.)

$$\text{The strain in the } x\text{-direction, } \epsilon_x = \frac{50 \times 10^6}{207 \times 10^9} - \frac{0 \cdot 3(-80 \times 10^6)}{207 \times 10^9}$$

$$= (0 \cdot 242 + 0 \cdot 116)10^{-3} = 0 \cdot 358 \times 10^{-3}$$

therefore change in length $= 0 \cdot 358 \times 10^{-3} \times 120 = 0 \cdot 043$ mm
therefore new length of plate $= 120 \cdot 043$ mm

$$\text{The strain in the } y\text{-direction, } \epsilon_y = \frac{-80 \times 10^6}{207 \times 10^9} - \frac{0 \cdot 3 \times 50 \times 10^6}{207 \times 10^9}$$

$$= (-0 \cdot 386 - 0 \cdot 072)10^{-3}$$
$$= -0 \times 458 \times 10^{-3}$$

therefore change in length $= -0 \cdot 458 \times 10^{-3} \times 40 = -0 \cdot 018$ mm
therefore new width of plate $= 39 \cdot 982$ mm
new dimensions of plate are $120 \cdot 043$ mm long $\times$ $39 \cdot 982$ mm wide.

2.7   Volumetric strain

In the preceding section we introduced the idea that a solid could be subjected to stresses in three mutually perpendicular directions which would result in strains in these directions. Consider now a rectangular solid of sides a, b and c under the action of direct stresses σ_1, σ_2 and σ_3 (Fig. 2.12).

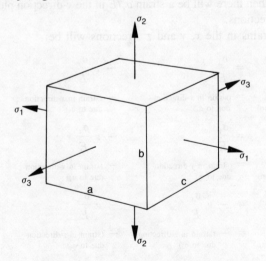

Fig. 2.12

The direct stresses will cause direct strains ϵ_1, ϵ_2 and ϵ_3 in the respective directions and the new dimensions of the solid will be a + ϵ_1a, b + ϵ_2b and c + ϵ_3c. The change in volume divided by the original volume is called the **volumetric strain**, where

$$\text{Volumetric strain} = \frac{(a + \epsilon_1 a)(b + \epsilon_2 b)(c + \epsilon_3 c) - abc}{abc}$$

$$= (1 + \epsilon_1)(1 + \epsilon_2)(1 + \epsilon_3) - 1$$
$$= 1 + \epsilon_1 + \epsilon_2 + \epsilon_3 + \epsilon_1\epsilon_2 + \epsilon_2\epsilon_3 + \epsilon_3\epsilon_1 + \epsilon_1\epsilon_2\epsilon_3 - 1$$
$$= \epsilon_1 + \epsilon_2 + \epsilon_3$$

since the strains are small and the products of small quantities can be considered negligible.

Converting these strains to stresses using Poisson's ratio gives:

$$\text{Volumetric strain} = \frac{\sigma_1 - \nu\sigma_2 - \nu\sigma_3}{E} + \frac{\sigma_2 - \nu\sigma_1 - \nu\sigma_3}{E} + \frac{\sigma_3 - \nu\sigma_1 - \nu\sigma_2}{E}$$

$$= \frac{(\sigma_1 + \sigma_2 + \sigma_3)(1 - 2\nu)}{E}$$

For a cube with $\sigma_1 = \sigma_2 = \sigma_3 = \sigma$ this reduces to:

$$\text{Volumetric strain} = \frac{3\sigma(1 - 2\nu)}{E} \qquad (2.14)$$

Thus a three-dimensional stress situation produces a three-dimensional strain or volumetric strain. The ratio of σ, the direct stress, to the volumetric strain is termed the **bulk modulus** for the material, K.

$$\text{Thus, } K = \frac{\sigma}{3\sigma(1 - 2\nu)/E}$$

$$\text{or } K = \frac{E}{3(1 - 2\nu)} \text{ N/m}^2 \qquad (2.15)$$

An important application of bulk modulus is when a body is subjected to hydrostatic pressure, for it is then that the applied stresses are equal in all directions. Then K is defined as the ratio of hydrostatic pressure to volumetric strain.

Problems 2.4

1 Calculate the change in diameter of a steel bar subjected to a tensile force of 15 kN, given that the original diameter was 25 mm, E for steel is 207 GN/m^2 and ν for steel is 0·3.

2 A steel bar of 50 mm square section and 1·2 m long is subjected to an axial tensile load of 400 kN. Determine the decrease in the lateral dimensions if $E = 208$ GN/m^2 and $\nu = 0·3$.

3 A round steel bar, 25 mm diameter and 250 mm long, is loaded by an axial compressive force of 2×10^5 newtons. Calculate the change in volume of the bar if $E = 200$ GN/m^2 and $\nu = 0.32$.

4 A bar of metal subjected to an axial compressive stress was measured and it was found that the change in length was 0.05% of its original length, while the change in its diameter was 0.016% of its original value. Calculate the value of Poisson's ratio for the metal of the bar.

5 A rectangular steel plate 100 mm × 50 mm is subjected to a tensile stress of 40 MN/m^2 on the longer side and a tensile stress of 25 MN/m^2 on the shorter side. Calculate the increase in area of the plate due to the applied stresses. (Take $E = 207$ GN/m^2 and $\nu = 0.3$)

6 A square steel plate of side 100 mm is subjected to a tensile stress of 50 MN/m^2 on one pair of parallel edges, and a tensile stress of 80 MN/m^2 on the parallel edges adjacent to the other two. Calculate the change in the area of the plate as a result of the applied stresses. Take $E = 206 \times 10^9$ N/m and $\theta = 0.29$.

7 If, in question 6, the 80 MN/m^2 stress were compressive, calculate then the change in area of the plate.

8 A mild steel component of section 10 mm square, and length 100 mm is subjected to an axial direct tensile load of 20 kN, a lateral tensile stress of 2 MN/m^2 on one face, and a lateral compressive stress of 1.75 MN/m^2 on the other. Calculate the change in volume of the component.

9 At a point on the surface of a pressurised copper pipe the axial stress is found to be 120 MN/m^2 tensile and the circumferential (hoop) stress (perpendicular to the axial stress) 160 MN/m^2 tensile. Calculate the strains in the directions of the applied stresses. ($E_{copper} = 110$ GN/m^2, $\nu_{copper} = 0.32$)

2.8 Simple bending theory

A beam is usually a horizontal structure designed to support vertical loads. The cross-sectional shape of the beam may be as any shown in Fig. 2.13, but we shall only concern ourselves with those in which the cross-sectional shape is symmetrical about a vertical axis.

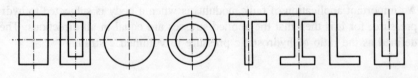

Fig. 2.13 Cross-sectional shapes of beams (The L-shape is not symmetrical about the vertical axis.)

BEAM SUPPORTS

The beam may be supported in one of two main ways, either simply (Fig. 2.14a) or by cantilever (Fig. 2.14b). In the simply supported case the beam is supported by vertical reactions, usually, though not always, at the extreme ends of the beam. In the cantilever case the beam is built in at one end, the fixing wall providing a vertical reaction and a fixing moment.

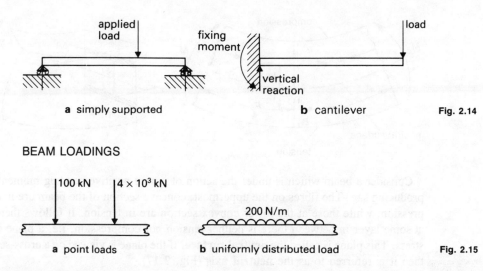

a simply supported   **b** cantilever   Fig. 2.14

BEAM LOADINGS

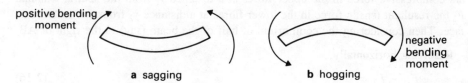

a point loads   **b** uniformly distributed load   Fig. 2.15

The type of loading that might be applied to the beam could be point (or concentrated) loading, and/or uniformly distributed loading (Fig. 2.15a and b) in which the total load is spread evenly over the beam so that each unit length has the same loading. Under the action of the applied loading the beam will deflect. The deflection may be in the form of a sagging distortion (Fig. 2.16a) or a hogging distortion (Fig. 2.16b).

positive bending moment   negative bending moment

a sagging   **b** hogging   Fig. 2.16

A beam is caused to sag by a load system which in effect, results in a moment tending to turn up the ends of the beam. This **bending moment** is conventionally given a positive sign, so that a positive bending moment gives a *sagging* beam. Similarly a beam is caused to hog by a negative bending moment which tends to turn the ends of the beam down, i.e. a negative bending moment gives a *hogging* beam.

THE BENDING OF A BEAM

When considering the bending of a beam, the following main assumptions are made.

a The beam is initially straight before the load is applied.
b Sections of the beam that are perpendicular to the axis remain perpendicular during bending.
c The material obeys Hooke's law, and the limit of elasticity is not exceeded.
d Young's modulus of elasticity has the same value for tension and compression.
e Every layer of material is free to expand or contract longitudinally and laterally under stress as if separate from other layers.
f The cross-section of the beam is symmetrical about the plane of bending, i.e. no twisting takes place.

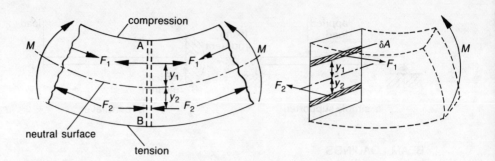

compression

M

A

F_1 F_1 M δA M

y_1

F_2 F_2

F_1

y_1

F_2

y_2

B

neutral surface

tension

Fig. 2.17

Consider a beam which is under the action of a pure positive bending moment, M, producing sag. The fibres on the uppermost, concave section of the beam are in compression, while those at the lower convex section are in tension. It follows then that at some layer in between there is neither tension nor compression, i.e. a plane of no stress. This plane is called the **neutral surface**. If the plane is viewed on a cross-section then it is referred to as the **neutral axis** (Fig. 2.17).

MOMENT OF RESISTANCE

The fibres above the neutral surface are in compression and those below the neutral surface are in tension. Consider a cut section of the beam AB. Suppose F_1 is the resultant compressive force in the upper fibres at a distance y_1 from the neutral axis and F_2 the resultant tensile force in the lower fibres at a distance y_2 from the neutral surface. Then considering the equilibrium of half of the beam (say) to the right of AB,

Resolving horizontally,

$$F_1 + F_2 = 0 \tag{2.16}$$

Taking moments about the neutral axis,

$$F_1 y_1 + F_2 y_2 = M \tag{2.17}$$

The internal moment $F_1 y_1 + F_2 y_2$ is termed the **moment of resistance** and is the material's resistance to the applied bending moment M. The moment of resistance of the section will increase equally with an increasing applied bending moment, up to a maximum value. Further increase of applied bending moment beyond the maximum moment of resistance of the material will result in failure of the beam.

THE STRESS DUE TO BENDING

Consider a section of a beam abcd which is subject to an applied positive bending moment M as shown in Fig. 2.18. If ef is the neutral axis then its length has remained unchanged from its value before bending. The length ab however is now compressed and the length cd is extended from their original lengths which *were* equal to ef.

Consider the stretched fibres in the convex portion of the beam cd.

Increase in length = cd − ef

therefore   strain = $\dfrac{cd - ef}{ef}$

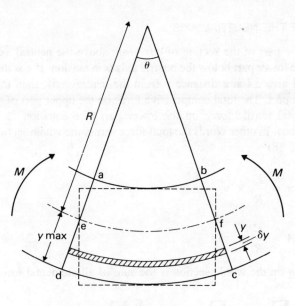

Fig. 2.18

but   cd $= (R + y)\theta$
and   ef $= R\theta$.

where θ is in radians, and R is the radius of curvature.

Therefore, strain $= \dfrac{y}{R}$

Also the strain in the material $= \dfrac{\sigma}{E}$ where σ is the stress at a distance y from the neutral axis.

therefore $\dfrac{y}{R} = \dfrac{\sigma}{E}$

or $\dfrac{\sigma}{y} = \dfrac{E}{R}$ \hfill (2.18)

Since both E and R are constants for a beam at a particular section, then equation (2.18) shows that the stress in a material is proportional to the distance of the stress from the neutral axis. The stress distribution on a cut section of the beam is thus linear and looks like Fig. 2.19.

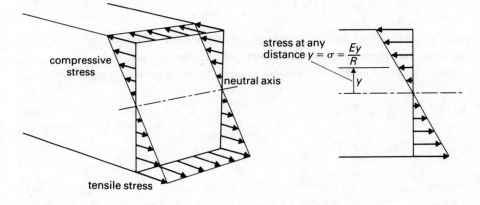

compressive stress

stress at any distance $y = \sigma = \dfrac{Ey}{R}$

neutral axis

tensile stress

Fig. 2.19

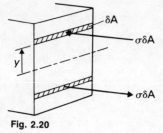

Fig. 2.20

POSITION OF THE NEUTRAL AXIS

We have shown that the part of the section of our beam above the neutral axis is in compression, whilst the lower part below the neutral axis is in tension. If σ is the stress acting on an elemental area δA at a distance y from the neutral axis, then the force acting on the area δA is $\sigma \delta A$. The total compressive force on the upper part of the section must equal the total tensile force on the lower part (see equation (2.16) for equilibrium of the section). In other words the total force across the whole section must be zero. Now, from (2.18)

$$\sigma = \frac{E}{R} \times y$$

therefore $\quad \sigma \delta A = \dfrac{E}{R} \times y \delta A$

The total force acting on the whole section is the sum of all elemental forces, i.e.

$$\text{total force} = \sum \sigma \delta A = \sum \frac{E}{R} y \delta A = \frac{E}{R} \sum y \delta A$$

In the limit as $\delta A \to 0$

$$\text{total force} = \int \sigma dA = \frac{E}{R} \int y dA$$

If the total force is to be zero, then $\int y dA$ must be zero, since clearly E/R cannot be. $\int y dA$ is the first moment of area of the section about the neutral axis (see Appendix 1), and so for this to be zero, the *neutral axis must pass through the centroid of the section*.

The stress distribution across the face of a section is linear. If σ is the stress at a distance y from the neutral axis then the stress at a distance x will be $x/y.\sigma$.

The force on an elemental area δA at a distance x from the neutral axis will be $x/y.\sigma.\delta A$.

The moment of the force about the neutral axis will be $x/y \times \sigma \delta A \times x$.

The total moment of resistance will then be given by

$$\int \frac{x^2 \sigma}{y} \, dA = \frac{\sigma}{y} \int x^2 dA$$

Now, $\int x^2 dA$ is termed the second moment of area of the section and is given the symbol I (see Appendix 2), thus total moment of resistance $= \sigma/y \times I$.

Since the total moment of resistance is equal to the applied bending moment M,

$$M = \frac{\sigma}{y} \times I$$

or $\quad \dfrac{M}{I} = \dfrac{\sigma}{y}$ 　　　　　　　　　　　　　　　　　　(2.19)

This formula gives the stress σ at any distance y from the neutral axis in terms of M and I. Combining equations (2.18) and (2.19) gives

$$\frac{M}{I} = \frac{\sigma}{y} = \frac{E}{R} \qquad (2.20)$$

where $M =$ applied bending moment in Nm

$\quad I =$ second moment of area about neutral axis in m^4

$\quad E =$ Young's modulus of elasticity in N/m^2

$\quad R =$ radius of curvature of neutral axis in m

$\quad y =$ distance from neutral axis to portion of beam being considered (normally the outer edges where stresses are a maximum) in m

$\quad \sigma =$ stress (tensile or compressive) at a distance y from the neutral axis in N/m^2

Note: Quantities should be *all* in m or *all* in mm. Equation (2.20) is the fundamental equation for the bending of beams and can be used to solve many problems at this level.

Example 2.9

A vertical steel post 3 m high has an external diameter of 100 mm and an internal diameter of 75 mm. Find the maximum stress due to bending when a horizontal force of 2 kN is applied at the top of the post.

The maximum stress due to bending will occur at the base of the post where the bending moment is a maximum. Thus

$$\text{maximum bending moment} = 2 \times 10^3 \times 3 = 6000 \text{ Nm}$$

$$I \text{ for an annular cross-section} = \frac{\pi(D^4 - d^4)}{64} \text{ (see Appendix 2)}$$

$$= \frac{\pi(0 \cdot 1^4 - 0 \cdot 075^4)}{64} = 3 \cdot 356 \times 10^{-6} \text{ m}^4$$

Applying $\dfrac{M}{I} = \dfrac{\sigma_b}{y}$, $\sigma_b = \dfrac{My}{I} = \dfrac{6000 \times 0 \cdot 050}{3 \cdot 356 \times 10^{-6}} = 89 \cdot 4 \text{ MN/m}^2$

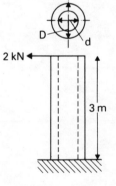

Example 2.10

A piece of flat steel has to be bent round a drum 2 m in diameter. What is the maximum thickness the strip can be made so that there shall be no permanent deformation when it is removed from the drum? The elastic limit of the steel is 216 MN/m^2 and $E = 216$ GN/m^2.

Let maximum thickness be t.

$\quad \sigma_{\text{max}} = 216 \times 10^6 \text{ N/m}^2$ acting at surface

$\quad E = 216 \times 10^9 \text{ N/m}^2$

$\quad R = 1\text{m}$

$\quad y = t/1$ since neutral axis is through the centroid of the section which is at the mid point

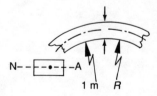

Applying $\dfrac{\sigma}{y} = \dfrac{E}{R}$, $y = \dfrac{\sigma}{E} \times R$

or $\quad t = \dfrac{2\sigma R}{E} = \dfrac{2 \times 216 \times 10^6 \times 1}{216 \times 10^9} = 2$ mm

Example 2.11

A girder of I-section has a depth of 250 mm and a second moment area of 10^8 mm^4. It is 6 m long, simply supported at its ends and carries a uniformly distributed load of 30 kN per metre. Calculate:

a the maximum bending moment
b the maximum stress in the girder
c the radius of curvature at the point of greatest bending moment

Take $E = 205$ GN/m^2.

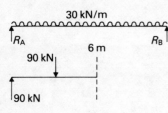

Because of the symmetry of the girder, $R_A = R_B = 90$ kN.
Maximum bending moment occurs at centre. Considering only $\frac{1}{2}$ beam, and replacing the distributed load of 30×3 kN by a point load of 90 kN in the centre:

a Max. bending moment $= 90 \times 3 - 90 \times 1\frac{1}{2} = 135$ kNm

b Using $\dfrac{M}{I} = \dfrac{\sigma}{y}$, $\sigma_{max} = \dfrac{My_{max}}{I} = \dfrac{135 \times 10^3 \times 0 \cdot 125}{10^8 \times 10^{-12}} = 168 \cdot 75$ MN/m^2

c Using $\dfrac{M}{I} = \dfrac{E}{R}$, $R = \dfrac{EI}{M} = \dfrac{205 \times 10^9 \times 10^8 \times 10^{-12}}{135 \times 10^3} = 151 \cdot 85$ m

---

Problems 2.5

1 A steel beam of rectangular section 600 mm deep can carry a uniformly distributed load, including its own weight, not greater than 270 kN when used as a simply supported beam over a span of 12 m. Calculate:
  a the stress due to bending when the beam is carrying the maximum load, if the second moment of area of the section is $1 \cdot 03 \times 10^{-3}$ m^4
  b the radius of curvature at the point of maximum bending moment if $E = 193$ GN/m^2

2 a A steel bar has a rectangular section 200 mm by 50 mm. If the bar is used as a beam, compare the relative strengths in resisting bending moment when it is:
  i laid flat
  ii laid on the smaller edge
  b If when laid flat the bar rests on supports 2 m apart, find the uniformly distributed load to give a maximum stress due to bending of 57 MN/m^2

3 A length of solid circular section bar is to be bent into the arc of a circle 3 m diameter but must be capable of recovering its straight form. Determine the maximum diameter for the cross-section of the bar if the elastic limit stress is 232 MN/m^2 and $E = 201$ GN/m^2. What will be the maximum bending moment exerted by the beam in the bent form?

4 A round bar, 125 mm diameter, is to be used as a beam. Find the maximum allowable bending moment if the stress due to bending is limited to $17 \cdot 21$ MN/m^2. Calculate also the radius of curvature at the point of maximum bending moment if $E = 193$ GN/m^2.

5 An aircraft's tail plane spar is 150 mm deep at the section of maximum bending moment, its cross section is symmetrical about the neutral axis (NA) and $I_{NA} = 2 \times 10^6$ mm^4.
  a Calculate the maximum bending moment which the spar can withstand if the maximum compressive stress is limited to 155 MN/m^2.
  b Regarding the spar as a cantilever $1 \cdot 3$ m long, what load, uniformly distributed over the whole length, would produce this bending moment?

6 Calculate the wall thickness required for a tube 50 mm outside diameter, which has to carry a bending moment of 650 Nm with a maximum allowable compressive stress of 324 MN/m^2.

SECTION MODULUS

From equations (2.19) $\dfrac{M}{I} = \dfrac{\sigma}{y}$

When $y = y_{max}$ then $\sigma = \sigma_{max}$

therefore $\dfrac{M}{I} = \dfrac{\sigma_{max}}{y_{max}}$ or $M = \sigma_{max} \times \dfrac{I}{y_{max}}$.

The term $\dfrac{I}{y_{max}}$ is termed the **section modulus**, (or elastic modulus), Z

$$M = \sigma_{max} \times Z \tag{2.21}$$

In the case of a rectangular cross-section breadth b and depth D

$$I_{NA} = \frac{bD^3}{12} \quad \text{therefore} \quad Z_{NA} = \frac{bD^2}{6}$$

For a circular cross-section of diameter d

$$I_{NA} = \frac{\pi d^4}{64}, \quad \text{therefore} \quad Z_{NA} = \frac{\pi d^3}{32}$$

For the various profile sections in commercial use, I beams, channels etc., the magnitudes of I_{NA} and Z for the standard manufactured sizes are tabulated in handbooks. An abridged list of such sections follows; a full list can be found in BS 4, Part 1 (1980). The procedure is as follows:

1 Calculate the maximum bending moment in Nmm.
2 Decide on maximum allowable stress in beam (a maximum permissible stress in bending of 165 N/mm^2 is usual).
3 Calculate Z in mm, and convert to cm by dividing by 1000.

Table 2.1

UNIVERSAL BEAMS
To: BS4 Part 1

Designation		Depth of section D	Width of section B	Thickness		Area of section	Moment of inertia		Radius of gyration		Elastic modulus		Plastic modulus	
Serial size	Mass per metre			Web t	Flange T		Axis x·x	Axis v·v	Axis x·x	Axis v·v	Axis x·x	Axis v·v	Axis x·x	Axis v·v
mm	kg	mm	mm	mm	mm	cm²	cm²	cm⁴	cm	cm	cm³	cm³	cm³	cm³
914 × 419	388	920·5	420·5	21·6	36·6	494·5	718742	45407	38·1	9·58	15616	2160	17657	3339
	343	911·4	418·5	19·4	32·0	437·6	625282	39150	37·8	9·46	13722	1871	15474	2890
914 × 305	289	926·6	307·8	19·6	32·0	368·8	504594	15610	37·0	6·51	10891	1014	12583	1603
	253	918·5	305·6	17·3	27·9	322·8	436610	13318	36·8	6·42	9507	871·9	10947	1372
	224	910·3	304·1	15·9	23·9	285·3	375924	11223	36·3	6·27	8259	738·1	9522	1162
	201	903·0	303·4	16·2	20·2	256·4	325529	9427	35·6	6·06	7210	621·4	8362	962·5
838 × 292	226	850·9	293·8	16·1	26·8	288·7	339747	11353	34·3	6·27	7986	772·9	9157	1211
	194	840·7	292·4	14·7	21·7	247·2	279450	9069	33·6	6·06	6648	620·4	7648	974·4
	176	834·9	291·6	14·0	18·8	224·1	246029	7792	33·1	6·90	5894	534·4	6809	841·6
762 × 267	197	769·6	268·0	15·6	25·4	250·8	239694	6174	30·9	5·71	6234	610·0	7167	958·7
	173	762·0	266·7	14·3	21·6	220·6	205177	6846	30·5	5·57	5385	613·4	6197	807·3
	147	753·9	265·3	12·9	17·5	188·1	168966	5468	30·0	5·39	4483	412·3	5174	649·0
686 × 254	170	692·9	255·8	14·5	23·7	216·6	170147	6621	28·0	5·53	4911	517·7	5624	810·3
	152	687·6	254·5	13·2	21·0	193·8	150319	5782	27·8	5·46	4372	454·5	4997	710·0
	140	683·6	263·7	12·4	19·0	178·6	136276	5179	27·6	5·38	3988	408·2	4560	637·8
	125	677·9	253·0	11·7	16·2	159·6	118003	4397	27·2	5·24	3481	346·1	3996	542·0
610 × 305	238	633·0	311·5	18·6	31·4	303·8	207571	15838	26·1	7·22	6559	1017	7456	1574
	179	617·5	307·0	14·1	23·6	227·9	161631	11412	25·8	7·08	4911	743·3	5521	1144
	149	609·6	304·8	11·9	19·7	190·1	124660	9300	25·6	6·99	4090	610·3	4572	936·8
610 × 229	140	617·0	230·1	13·1	22·1	178·4	111844	4512	25·0	5·03	3626	392·1	4146	612·6
	125	611·9	229·0	11·9	19·6	159·6	98579	3933	24·7	4·96	3222	343·5	3677	535·7
	113	607·3	228·2	11·2	17·3	144·5	87431	3439	24·6	4·88	2879	301·4	3288	470·2
	101	602·2	227·6	10·6	14·8	129·2	75720	2912	24·2	4·75	2515	255·9	2882	400·0
533 × 210	122	544·6	211·9	12·8	21·3	155·8	76207	3393	22·1	4·67	2799	320·2	3203	500·6
	109	539·5	210·7	11·6	18·8	138·6	86739	2937	21·9	4·60	2474	278·8	2824	435·1
	101	536·7	210·1	10·9	17·4	129·3	61659	2694	21·8	4·56	2296	256·5	2620	400·0
	92	533·1	209·3	10·2	15·6	117·8	66363	2392	21·7	4·51	2076	228·6	2366	356·2
	82	528·3	208·7	9·6	13·2	104·4	47491	2006	21·3	4·38	1798	192·2	2056	300·1
457 × 191	98	467·4	192·8	11·4	19·6	125·3	45717	2343	19·1	4·33	1956	243·0	2232	378·3
	89	463·6	192·0	10·6	17·7	113·9	41021	2086	19·0	4·28	1770	217·4	2014	337·9
	82	460·2	191·3	9·9	16·0	104·5	37103	1871	18·8	4·23	1612	195·6	1833	304·0
	74	457·2	190·5	9·1	14·5	95·0	33388	1671	18·7	4·19	1461	175·5	1657	272·2
	67	453·6	189·9	8·5	12·7	85·4	29401	1452	18·5	4·12	1296	152·9	1471	237·3
457 × 152	82	465·1	153·5	10·7	18·9	104·5	36215	1143	18·6	3·31	1557	149·0	1800	235·4
	74	461·3	152·7	9·9	17·0	95·0	32435	1012	18·5	3·26	1406	132·5	1622	209·1

Table 2.1

UNIVERSAL BEAMS
To: BS4 Part 1

Designation		Depth of section D	Width of section B	Thickness		Area of section	Moment of inertia		Radius of gyration		Elastic modulus		Plastic modulus	
Serial size	Mass per metre			Web t	Flange T		Axis x·x	Axis v·v	Axis x·x	Axis v·v	Axis x·x	Axis v·v	Axis x·x	Axis v·v
mm	kg	mm	mm	mm	mm	cm²	cm²	cm⁴	cm	cm	cm³	cm³	cm³	cm³
	67	457·2	151·9	9·1	15·0	85·4	28577	878	18·3	3·21	1250	115·5	1441	182·2
	60	454·7	152·9	8·0	13·3	75·9	25464	794	18·3	3·23	1120	103·9	1284	162·9
	52	449·8	152·4	7·6	10·9	66·5	21345	645	17·9	3·11	949·0	84·6	1094	133·2
406 × 178	74	412·8	179·7	9·7	16·0	95·0	27329	1545	17·0	4·03	1324	172·0	1504	266·9
	67	409·4	178·8	8·8	14·3	85·5	24329	1365	16·9	4·00	1188	152·7	1346	236·5
	60	406·4	177·8	7·8	12·6	76·0	21508	1199	16·8	3·97	1058	134·8	1194	208·3
	54	402·6	177·6	7·6	10·9	68·4	18626	1017	16·5	3·85	925·3	114·5	1048	177·5
406 × 140	46	402·3	142·4	6·9	11·2	59·0	15647	539	16·3	3·02	777·8	75·7	888·4	118·3
	39	397·3	141·8	6·3	8·6	49·4	12452	411	15·9	2·89	626·9	58·0	720·8	91·08
356 × 171	67	364·0	173·2	9·1	15·7	85·4	19522	1362	15·1	3·99	1073	157·3	1212	243·0
	57	358·6	172·1	8·0	13·0	72·2	16077	1109	14·9	3·92	896·5	128·9	1009	198·8
	51	355·6	171·5	7·3	11·5	64·6	14156	968	14·8	3·87	796·2	112·9	894·9	174·1
	45	352·0	171·0	6·9	9·7	57·0	12091	812	14·6	3·78	686·9	95·0	773·7	146·7
356 × 127	39	352·8	126·0	6·5	10·7	49·4	10087	357	14·3	2·09	571·8	56·6	653·6	88·68
	33	348·5	125·4	6·9	8·5	41·8	8200	280	14·0	2·59	470·6	44·7	539·8	70·24
305 × 165	54	310·9	166·8	7·7	13·7	68·4	11710	1061	13·1	3·94	753·3	127·3	844·8	195·3
	46	307·1	165·7	6·7	11·8	58·9	9948	897	13·0	3·90	647·9	108·3	722·7	165·8
	40	303·8	165·1	6·1	10·2	51·5	8523	763	12·9	3·85	561·2	92·4	624·5	141·5
305 × 127	48	310·4	125·2	8·9	14·0	60·8	9504	460	12·5	2·75	612·4	73·5	706·1	115·7
	42	306·6	124·3	8·0	12·1	53·2	8143	388	12·4	2·70	531·2	62·5	610·5	98·24
	37	303·8	123·5	7·2	10·7	47·5	7162	337	12·3	2·67	471·5	54·6	540·5	85·66
305 × 102	33	312·7	102·4	6·6	10·8	41·8	6487	193	12·5	2·15	415·0	37·8	479·9	59·85
	28	306·9	101·9	6·1	8·9	36·3	5421	157	12·2	2·08	351·0	30·8	407·2	48·92
	25	304·8	101·6	5·8	6·8	31·4	4387	120	11·8	1·96	287·9	23·6	337·8	37·98
254 × 146	43	259·6	147·3	7·3	12·7	55·1	6558	677	10·9	3·51	505·3	92·0	568·2	141·2
	37	256·0	146·4	6·4	10·9	47·5	5556	571	10·8	3·47	434·0	78·1	485·3	119·6
	31	251·5	146·1	6·1	8·6	40·0	4439	449	10·5	3·35	353·1	61·5	395·6	94·52
254 × 102	28	260·4	102·1	6·4	10·0	36·2	4008	178	10·5	2·22	307·9	34·9	353·4	54·84
	25	257·0	101·9	6·1	8·4	32·2	3408	148	10·3	2·14	265·2	29·0	305·6	45·82
	22	254·0	101·6	5·8	6·8	28·4	2867	120	10·0	2·05	225·7	23·6	261·9	37·55
203 × 133	30	206·8	133·8	6·3	9·6	38·0	2887	384	8·72	3·18	279·3	57·4	131·3	88·05
	25	203·2	133·4	5·8	7·8	32·3	2358	310	8·54	3·10	231·9	46·4	259·8	71·39

Reproduced by kind permission of the Director, The Steel Construction Institute, Ascot from *A check list for Designers — Structural Steel Sections*, from whom up-to-date copies can be obtained.

4 Determine value of Z in tables which is slightly greater than the calculated value to give the required section.

Example 2.12

A simply supported I-section beam of length $3 \cdot 5$ m is required to carry a concentrated load of 6×10^4 newtons at the centre of its span. If the maximum permissible stress allowable in the beam is 165 N/mm^2, select a suitable beam from standard tables using a calculated value of section modulus.

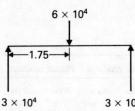

Maximum bending moment, $M = 3 \times 10^4 \times 1 \cdot 75 \times 10^3 = 5 \cdot 25 \times 10^7$ Nmm

From equation (2.21) $M = \sigma_{max} \times Z$

$$\text{therefore } Z = \frac{M}{\sigma_{max}}$$

$$= \frac{5 \cdot 25 \times 10^7}{165} = 3 \cdot 18 \times 10^5 \text{ mm}^3$$

$$= 318 \cdot 0 \text{ cm}^3$$

From Table 2.1, a beam with section modulus of just greater than 318 cm^3 is required.
By inspection a serial size of 305×102 mm is most suitable, giving a section modulus of 351 cm^3.

Problems 2.6

Using standard I-section tables calculate the most suitable serial size of beam for the following applications:

1 A simply supported beam of length 5 cm carrying a concentrated load of 10^3 newtons at the centre of its span if the maximum permissible stress is 165 N/mm^2.

2 A cantilever beam of length $2 \cdot 6$ m with a concentrated load at its free end of 400 kN, assuming a maximum permissible stress of 150 N/mm^2.

3 A simply supported beam of length 14 m carrying a uniformly distributed load of 20 kN/m along the whole of its span, assuming a maximum permissible stress of 175 N/mm^2.

2.9   Shear force and bending moment

Engineering structures and components always react to the external loads and forces imposed upon them. To begin with applied loads are always balanced by the forces which support or hold the structure. These external forces form an equilibrium state which responds to basic equilibrium analysis.

In addition engineers must ensure that the component and the material from which the component is made are strong enough and of adequate size to withstand the internal resistive forces that will arise within the component. This section attempts to describe the action of the external and internal forces acting on simple beam structures, and the methods of calculating their magnitudes.

CALCULATION OF SUPPORT REACTIONS

A beam is a horizontal structural member carrying external transverse loads. The external loads may be in the form of concentrated point loads which may include the weight of the beam and the reaction of supports and/or uniformly distributed loads (UDL) over part or all of the beam. The beam may be simply supported, with or without an overhang, or a cantilever.

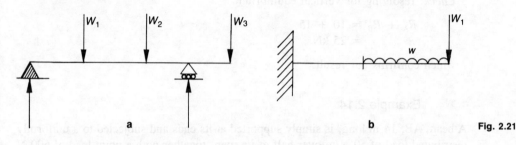

Fig. 2.21

A simply supported beam has two supports. One of the supports is in the form of a pivot and the other is carried on frictionless rollers. The rollers are incapable of sustaining any horizontal component of load and since we will confine ourselves to only vertically applied loads then it follows that for horizontal equilibrium the reactions of the supports must also be purely vertical. The pivot ensures that the net moment of the applied loads and the support reactions must be zero. There are thus two basic conditions of equilibrium that are required to be satisfied.

1 The net vertical force is zero.
2 The net turning moment about any point is zero.

For a simply supported beam with known applied loads there are thus two equations of equilibrium which are sufficient to determine the values of the two support reactions. In practice it is preferable to take moments twice about different points on the beam, and to reserve the equation for vertical equilibrium as a useful check of the accuracy of the calculation.

Example 2.13

A beam AB is simply supported at A and B and is 10 m long. It is subjected to concentrated loads of 10 kN and 15 kN at distances of 4 m and 8 m from the left-hand end. Calculate the reactions of the supports.

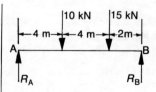
Fig. 2.22

Let the reactions at A and B be R_A and R_B kN respectively.

For equilibrium of the beam the net turning moments about any point is zero. In practice it is best to take moments about the points at which the reactions act since this will eliminate each reaction in turn from the equation. Thus:

taking moments about A
$$R_B \times 10 = 10 \times 4 + 15 \times 8$$

$$R_B = \frac{40 + 120}{10} = 16 \text{ kN}$$

and, taking moments about B
$$R_A \times 10 = 10 \times 6 + 15 \times 2$$

$$R_A = \frac{60 + 30}{10} = 9 \text{ kN}$$

Check: resolving for vertical equilibrium

$$R_A + R_B = 10 + 15$$
$$= 25 \text{ kN}$$

which confirms our result.

Example 2.14

A beam AB, 14 m long, is simply supported at its ends and subjected to a uniformly distributed load of 50 N/m over half of its span, together with a point load of 600 N in the centre of the remaining half. Calculate the reactions of the supports.

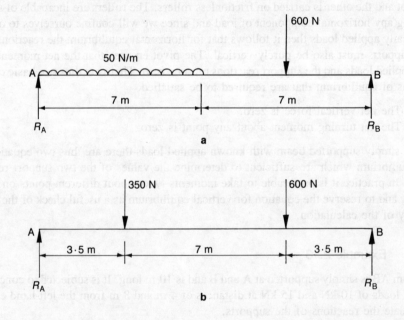

Fig. 2.23

For the purposes of calculating the reactions the UDL is regarded as a single-point load of value $50 \times 7 = 350$ N at the centre of the part of the beam carrying the UDL (Fig. 2.23b).

Again, for equilibrium of the beam, taking moments about A

$$14R_B = 350 \times 3{\cdot}5 + 600 \times 10{\cdot}5$$

$$\therefore R_B = \frac{1225 + 6300}{14} = \frac{7525}{14} = \underline{537{\cdot}5 \text{ N}}$$

and, taking moments about B

$$14R_A = 350 \times 10{\cdot}5 + 600 \times 3{\cdot}5$$

$$\therefore R_A = \frac{3675 + 2100}{14} = \frac{5775}{14} = \underline{412{\cdot}5 \text{ N}}$$

Check: resolving vertically

$$R_A + R_B = 350 + 600 = 950 \text{ N}$$
$$\text{and } 537{\cdot}5 + 412{\cdot}5 = 950 \text{ N}$$

A cantilever beam has one end rigidly built into a wall (Fig. 2.21b). Considering the equilibrium of this arrangement there are again two conditions to be met;

1 The net vertical force is zero.
2 The net turning moment about any point on the beam is zero.

Example 2.15

A cantilever beam 12 m long carries loads of 120 kN at the free end and 80 kN at a distance of 4 m from the free end. Calculate the reactions at the support.

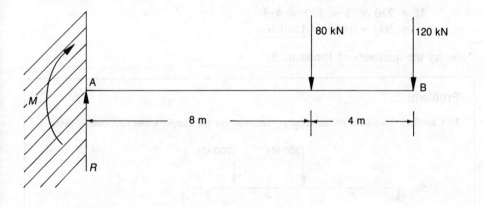

Fig. 2.24

For equilibrium of the beam it is clear that the built-in end must provide the necessary reactions. These reactions must be in the form of a vertical force R to give vertical equilibrium together with a clockwise moment M which provides rotational equilibrium.

Resolving vertically

$$R = 80 + 120$$
$$= 200 \text{ kN}$$

and, taking moments about A

$$M = 80 \times 8 + 120 \times 12$$
$$= 640 + 1440$$
$$= 2080 \text{ kNm}$$

Example 2.16

A cantilever beam AB, 6 m long, carries a UDL of 80 N/m over the half of the span furthest from the support. There is also a point load of 250 N at a distance of 4 m from the free end.

Calculate the reaction at the support.

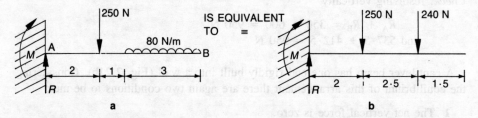

Fig. 2.25

As before the UDL of 80 N/m over 3 m is replaced by a point load of $80 \times 3 = 240$ N at the centre of the 3 m length (Fig. 2.25b).

For equilibrium of the beam, resolving vertically

$$R = 250 + 240 = \underline{490 \text{ N}}$$

and, taking moments about A

$$M = 250 \times 2 + 240 \times 4 \cdot 5$$
$$= 500 + 1080 = \underline{1580 \text{ Nm}}$$

Now try the questions of Problems 2.7.

Problems 2.7

1 Calculate the values of the support reactions for the simply supported beams shown.

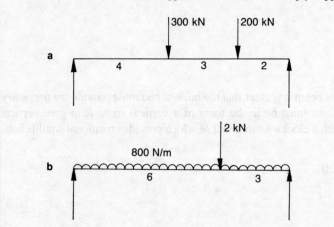

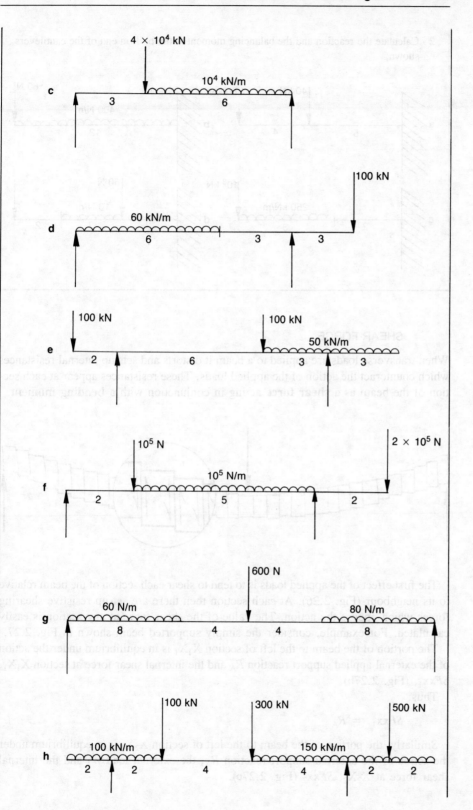

2 Calculate the reaction and the balancing moment at the built-in end of the cantilevers shown;

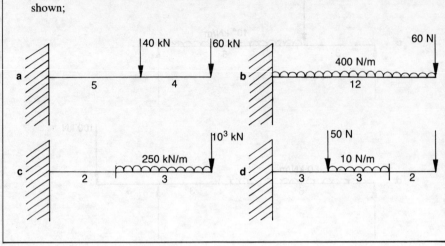

SHEAR FORCE

When transverse loads are applied to a beam it distorts and sets up internal resistances which counteract the action of the applied loads. These resistances appear at each section of the beam as a **shear force** acting in conjunction with a **bending moment**.

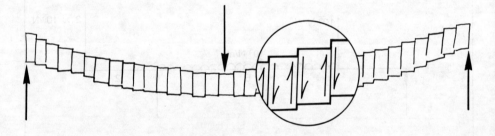

Fig. 2.26

The first effect of the applied loads is to tend to shear each section of the beam relative to its neighbour (Fig. 2.26). At each section then there are set up resistive shearing forces which oppose the action. The value of the shear force at any section is easily calculated. For example, consider the simply supported beam shown in Fig. 2.27.

The portion of the beam to the left of section X_1X_1 is in equilibrium under the action of the external applied support reaction R_A and the internal shear force at section X_1X_1, SF_{xx_1}, (Fig. 2.27b).

Thus;

$$SF_{xx_1}, = R_A$$

Similarly, the portion of the beam to the left of section X_2X_2 is in equilibrium under the action of the external support reaction R_A, the external load W_1 and the internal shear force at X_2X_2, SF_{xx_2} (Fig. 2.27c).

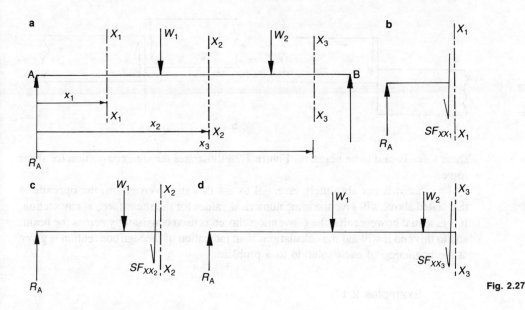

Fig. 2.27

Thus;

$$SF_{XX_2} = R_A - W_1$$

Again, considering the portion of the beam to the left of section X_3X_3, this is in equilibrium under the action of the external reaction R_A, the external loads W_1 and W_2 and the internal shear force at X_3X_3, SF_{XX_3} (Fig. 2.27d).
Thus;

$$SF_{XX_3} = R_A - W_1 - W_2$$

But resolving the external forces for the whole beam vertically gives

$$R_A - W_1 - W_2 = R_B$$

Thus $SF_{XX_3} = -R_B$

This is hardly surprising since this result would have been obtained had we considered the equilibrium of the portion of the beam to the *right* of section X_3X_3.

> In general we can say that the value of the shear force at any section of a beam is equal to the algebraic sum of the external applied loads, including the support reaction either to the left, or the right of the section.

SIGN CONVENTION FOR SHEAR FORCE

If, at a particular section of a beam, the shear forces are such as to tend to oppose the upward motion of the right-hand portion of the beam relative to the left then the shear force is said to be positive. If, however, the shear forces at a section are such as to oppose the downward motion of the right-hand side relative to the left then the

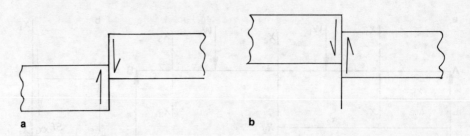

Fig. 2.28 a positive SF
b negative SF

a   b

shear force is said to be negative. Figure 2.28 illustrates the sign convention for shear force.

Note that it is not absolutely essential to use this sign convention, the opposite to that stated above will give the same numerical values for the shear force at any section. It is essential however that the convention chosen is used consistently across the beam, and to this end it will aid the calculations if an indication of the sign convention is given at the beginning of each solution to a problem.

Examples 2.17

Calculate the shear force at the sections X_1X_1, X_2X_2 and X_3X_3 for the simply supported beam shown:

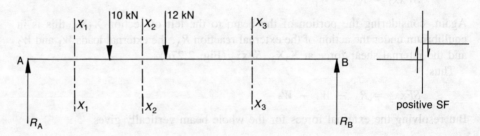

positive SF

For equilibrium of the beam as a whole it will be found that:

$$R_A = 13 \text{ kN} \quad \text{and} \quad R_B = 9 \text{ kN}$$

Considering the equilibrium of the portion of the beam to the left of section X_1X_1 we have:

$$SFxx_1 = -13 \text{ kN}$$

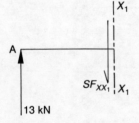

Note that SFxx is downwards resisting the upward motion of the left-hand side (LHS) of the beam relative to the right-hand side (RHS) and hence (RHS) and hence is negative in accordance with the sign convention indicated.

Considering the equilibrium of the portion of the beam to the left of section X_2X_2, we have:

$$SFxx_2 = -13 + 10$$
$$= -3 \text{ kN}$$

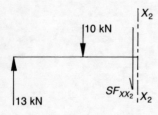

Sign convention as before with the 10 kN load supplying a SF at XX_2 in the opposite sense to that supplied by the reaction.

Considering the equilibrium of the portion of the beam to the left of section X_3X_3, we have:

$$SFxx_3 = -13 + 10 + 12$$
$$= +9 \text{ kN}$$

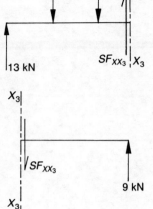

Here the arithmetic automatically ensures that $SFxx_3$ has the correct sign, i.e. SF opposes downward motion of left-hand side of beam relative to the right-hand side.

Note that if we had calculated SFxx by considering the portion of the beam to the *right* of XX we would have had:

$$SFxx_3 = +9 \text{ kN}$$

Sign convention as before: $SFxx_3$ opposes the upward motion of right-hand side of beam relative to the left-hand side.

This always provides a useful check that the calculations working from the left-hand side of the beam are correct.

Example 2.18

Calculate the shearing force at the sections shown on the beam AC carrying a UDL over the whole span and point load at E.

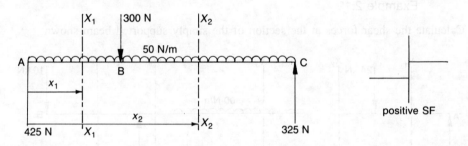

Considering the equilibrium of the beam as a whole it can be calculated that

$$R_A = 425 \text{ N} \quad \text{and} \quad R_B = 325 \text{ N}$$

Considering now the equilibrium of the portion of the beam to the left of XX_1. The UDL over this portion may be replaced by an equivalent point load. This load of $50\,x$ is placed in the centre of the portion.
Thus:

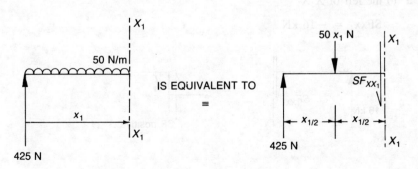

Then

$$SFxx_1 = -425 + 50x_1$$

Thus SF is a linear function of x_1, the distance of X_1X_1 from A.

Considering the equilibrium of the portion of the beam to the left of X_2X_2.

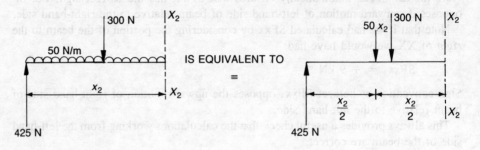

then

$$SFxx_2 = -425 + 300 + 50x_2$$

Again $SFxx_2$ is a linear function of x_2

Example 2.19

Calculate the shear forces at the section of the simply supported beam shown:

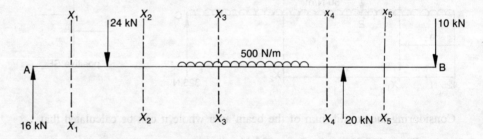

Considering the equilibrium of the beam as a whole gives the reactions as $R_A = 16$ kN and $R_B = 20$ kN.

Considering the equilibrium of the portion of the beam to the left of each of the sections in turn we have

a to the left of X_1X_1

$$SFxx_1 = -16 \text{ kN}$$

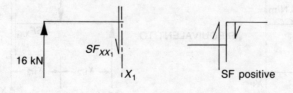

b to the left of X_2X_2

$$SFxx_2 = -16 + 24 = \underline{+8 \text{ kN}}$$

c to the left of X_3X_3

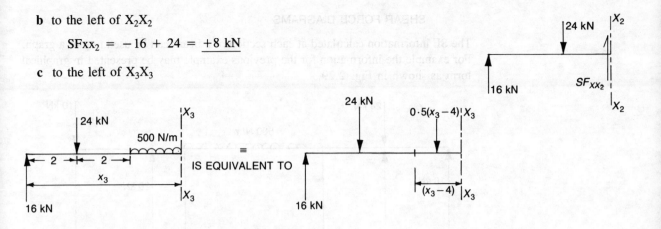

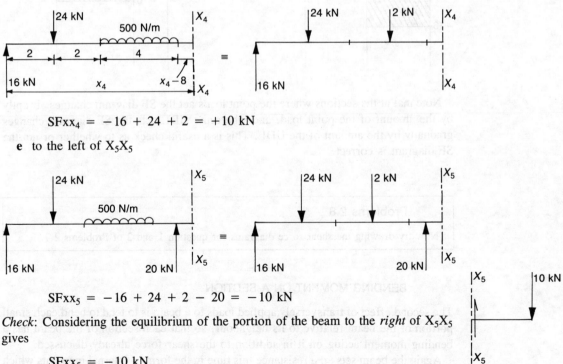

$$SFxx_3 = -16 + 24 + 0 \cdot 5 \ (x_3 - 4) \text{ kN}$$

Thus the shear force over the portion of the beam carrying the UDL depends on x_3 the distance of X_3X_3 from A.

When $x_3 = 4$ $SFxx_3 = -16 + 24 + 0 = +8 \text{ kN}$
and when $x_3 = 8$ $SFxx_3 = -16 + 24 + 2 = +10 \text{ kN}$

d to the left of X_4X_4

$$SFxx_4 = -16 + 24 + 2 = +10 \text{ kN}$$

e to the left of X_5X_5

$$SFxx_5 = -16 + 24 + 2 - 20 = -10 \text{ kN}$$

Check: Considering the equilibrium of the portion of the beam to the *right* of X_5X_5 gives

$$SFxx_5 = -10 \text{ kN}$$

SHEAR FORCE DIAGRAMS

The SF information calculated at each section of a beam can be used to plot a graph. For example the information for the previous example may be presented in graphical form as shown in Fig. 2.29.

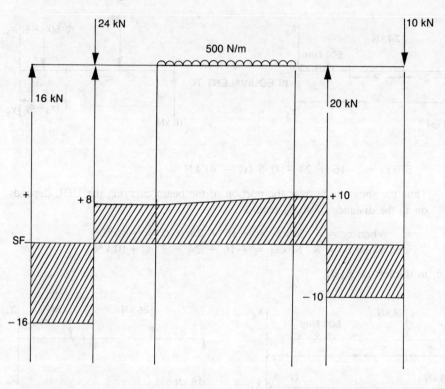

Fig. 2.29

Note that at the sections where the point loads act the SF diagram changes abruptly by the amount of the point load, and where the UDL acts the SF diagram changes gradually by the amount of the UDL. This is a useful check as to whether or not the SF diagram is correct.

Problems 2.8

Now try drawing the shear force diagrams for question 1 and 2 of Problems 2.7.

BENDING MOMENT ON A SECTION

The second effect of transversely applied loads to a beam is to tend to bend each small portion of the beam relative to its neighbour, so that each section of the beam has a bending moment acting on it in addition to the shear force already discussed.

Again the beam sets up a resistance this time in the form of internal moments which oppose the action of the applied loads. The value of the bending moment at any section

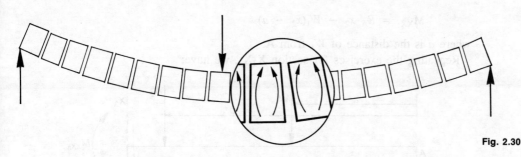

Fig. 2.30

may be calculated by considering the equilibrium of the total moments acting on the portion of the beam either to the left or right of the section.

Example 2.20

Consider the simply supported beam shown.

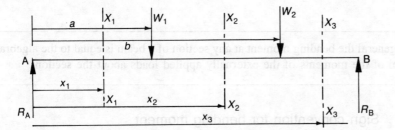

We have already seen that the portion of the beam to the left of X_1X_1 is in equilibrium under the action of the support reaction R_A and the shear force $SFxx_1$. For *complete* equilibrium however there must also be a moment applied at X_1X_1 to counteract the moment of R_A at that section.

If M_{XX_1} is the bending moment at X_1X_1 then, taking moments about X_1X_1

$$M_{XX_1} = R_A \, x_1$$

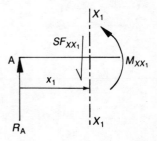

Similarly the portion of the beam to the left of section X_2X_2 is in equilibrium under the action of the support reaction R_A, the point load W_1 and the shear force $SFxx_2$. Taking moments about X_2X_2 we see that for complete equilibrium of the portion there is also a bending moment at X_2X_2 to counteract the moment effect of R_A and W_1 about that section.

If M_{XX_2} is the bending moment at X_2X_2, then taking moments about X_1X_1

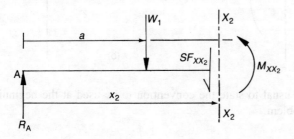

$$M_{XX_2} = R_A \cdot x_2 - W_1(x_2 - a)$$

where a is the distance of W_1 from A.

Repeating the exercises for section $X_3 X_3$, we have:

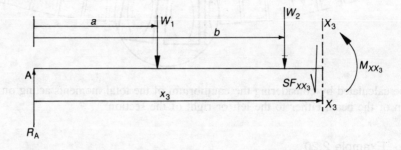

Taking moments about $X_3 X_3$ $M_{XX_3} = R_A \cdot x_3 - W_1(x_3 - a) - W_2(x_3 - b)$ where b is the distance of W_2 from A.

In general the bending moment at any section of a beam is equal to the algebraic sum of the moments of the externally applied loads about the section.

Sign convention for bending moment

To recap on section 2.8: a bending moment at a section that together with the applied load(s) causing it, tending to make that portion of the beam *sag* (i.e. the ends of the beam turn upwards) is regarded as a *positive* bending moment. A bending moment that tends to make the beam *hog* (i.e. the ends of the beam turn downwards) is regarded as a *negative* bending moment (Fig. 2.31).

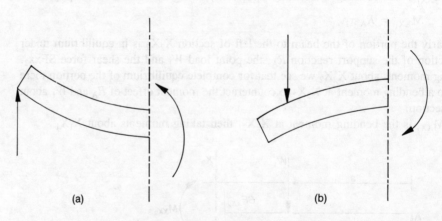

Fig. 2.31   (a)   (b)

In practice it is again usual to state the convention to be used at the beginning of the solution to each problem.

Example 2.21

Returning to the problem of Example 2.17 we can now calculate expressions for the bending moments at each of the sections X_1X_1, X_2X_2 and X_3X_3.

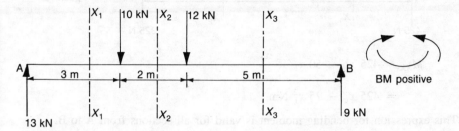

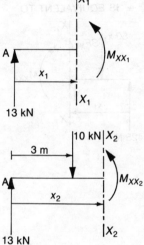

Consider the beam to the left of X_1X_1. Taking moments about X_1X_1

$$M_{XX_1} = 13x_1 \text{ kNm}$$

This expression for bending moment is valid for values of x from 0 to 3 m, i.e. up to the 10 kN load.

Consider the beam to the left of X_2X_2. Taking moments about X_2X_2

$$M_{XX_2} = 13x_2 - 10(x_2 - 3) \text{ kNm}$$

This expression for bending moment is valid for values of x_2 from 3 m to 5 m, i.e. the portion of the beam between the 10 kN and 12 kN loads.

For the beam to the left of X_3X_3

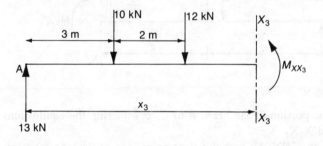

$$M_{XX_3} = 13x_3 - 10(x_3 - 3) - 12(x_3 - 5) \text{ kNm}$$

This expression is valid for values of x_3 from 5 m to 10 m, i.e. from the 12 kN load to the right-hand end.

Example 2.22

Given the simply supported beam ABC as in Example 2.18.

Consider the portion of the beam AB and in particular the portion to the left of section X_1X_1. For the purposes of considering equilibrium the UDL of 50 N/m may be replaced by a point load of 50 x_1 in the centre of the portion.

Taking moments about X_1X_1

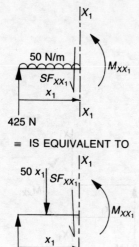

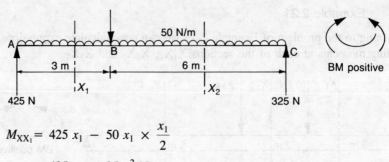

$$M_{XX_1} = 425\,x_1 - 50\,x_1 \times \frac{x_1}{2}$$

$$= 425\,x_1 - 25\,x_1^2 \text{ Nm}$$

This expression for bending moment is valid for all sections from A to B.

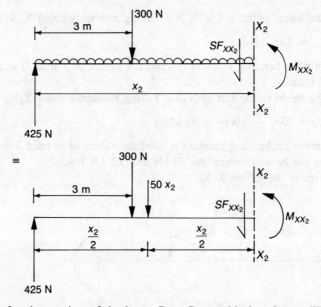

Similarly, for the portion of the beam B to C, considering the equilibrium of the beam to the left of X_2X_2.

Again replacing the UDL by its equivalent point load and taking moments about X_2X_2

$$Mxx_2 = 425\,x_2 - 300(x_2 - 3) - 50\,x_2$$

$$= 425\,x_2 - 300(x_3 - 3) - 25\,x_2^2 \text{ Nm}$$

This expression for bending moment is valid for all sections of the beam from B to C.

Example 2.23

Calculate the bending moments at the sections of the simply supported beam shown.

Considering the equilibrium of the portion of the beam to the left of each section in turn. (This time use a blank sheet of paper to cover the portions of the beam *not* under consideration.)

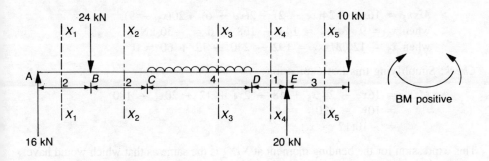

A to B

Considering portion of beam to the left of X_1X_1

$$Mxx_1 = 16\,x_1 \text{ kNm} \quad \text{when } x_1 = 0 \quad Mxx_1 = 0$$
$$\text{when } x_1 = 2 \quad Mxx_1 = 32 \text{ kNm}$$

B to C

for the beam to left of X_2X_2

$$Mxx_2 = 16\,x_2 - 24(x_2 - 2) \quad \text{when } x_2 = 2 \quad Mxx_2 = 32 \text{ kNm}$$
$$\text{when } x_2 = 4 \quad Mxx_2 = 16 \text{ kNm}$$

C to D

for the beam to the left of X_3X_3

$$Mxx_3 = 16\,x_3 - 24(x_3 - 2) - 0 \cdot 5(x_3 - 4)\,\frac{(x_3 - 4)}{2}$$

when $x_3 = 4$ $Mxx_3 = 16 \times 4 - 24 \times 2$ $= 16$ kN

$x_3 = 5$ $Mxx_3 = 16 \times 5 - 24 \times 3 - \dfrac{0 \cdot 5}{2}$ $= 7 \cdot 75$ kN

$x_3 = 6$ $Mxx_3 = 16 \times 6 - 24 \times 4 - 0 \cdot 5 \times 2 = -1$ kN
$x_3 = 7$ $Mxx_3 = 16 \times 7 - 24 \times 5 - 0 \cdot 5 \times \frac{9}{2} = -10 \cdot 25$ kN
$x_3 = 8$ $Mxx_3 = 16 \times 8 - 24 \times 6 - 0 \cdot 5 \times 8 = -20$ kN

D to E

for the beam to the left of X_4X_4

$$Mxx_4 = 16\,x_4 - 24(x_4 - 2) - 2(x_4 - 6)$$
$$\text{when } x_4 = 8 \quad Mxx_4 = 128 - 144 - 4 = -20 \text{ kNm}$$
$$\text{when } x_4 = 9 \quad Mxx_4 = 144 - 168 - 6 = -30 \text{ kNm}$$

E to F

for the beam to the left of X_5X_5

$$Mxx_5 = 16x_5 - 24(x_5 - 2) - 2(x_5 - 6) + 20(x_5 - 9)$$
when $x_5 = 9$ $Mxx_5 = 144 - 168 - 6 = -30$ kNm
when $x_5 = 12$ $Mxx_5 = 192 - 240 - 12 + 60 = 0$

Check: Simplifying this last expression

$$Mxx_5 = 16x_5 - 24 x_5 + 48 - 2x_5 + 12 + 20x_5 - 180$$
$$= 10x_5 - 120$$
$$= -10(12 - x_5)$$

This expression for the bending moment at X_5X_5 is the same as that which would have been obtained had we considered the portion of the beam to the *right* of X_5X_5.

BENDING MOMENT DIAGRAMS

As with shear force the values of bending moment may be illustrated in graphical form on a base of distance along the beam. Figure 2.32 shows the resulting bending moment diagram for the last worked example.

Note that for the portions of the beam carrying only point loads the bending moment diagram is a straight line, whereas the UDL results in a curved line. This is confirmed by reference to the expression for bending moment for each section.

The point at which the diagram changes from positive to negative is called a point of inflexion (or contraflexion). Although the shear force and bending moment diagrams

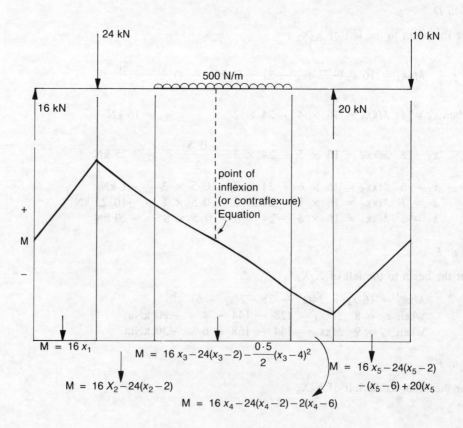

Fig. 2.32

have here been described and developed separately, it is more usual to see the two diagrams developed simultaneously.

Note that the relationship between loading shear force and bending moment is discussed in Chapter 7.

Example 2.24

A beam ABCDE is 16 m long and is simply supported at A and D 14 m apart. At B, 4 m from A, there is a concentrated load of 30 kN and a uniformly distributed load

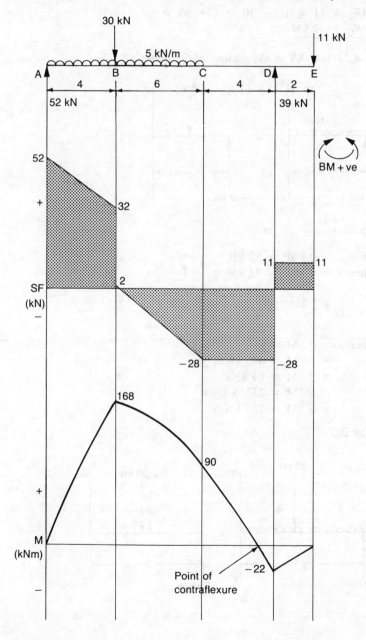

of 5 kN/m is carried over the portion AC which is 10 m long. A further concentrated load of 10 kN is carried at E on an overhang 2 m from D. Draw the shear force and bending moment diagrams indicating the maximum values on each.

For equilibrium of whole beam taking moments about D

$$14R_A = 30 \times 10 + 50 \times 9 - 11 \times 2$$
$$\underline{R_A = 52 \text{ kN}}$$

and, taking moments about A

$$14R_D = 11 \times 16 + 30 \times 4 + 50 \times 5$$
$$\underline{R_D = 39 \text{ kN}}$$

Consider a Section XX in the portion of the beam AB

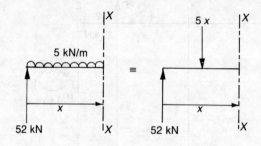

$$SF_{XX} = 52 - 5x$$

when $x = 0$ SF $= 52$ kN
when $x = 4$ SF $= 32$ kN

and $M_{XX} = 52x - \dfrac{5x^2}{2}$

when $x = 0$ M $= 0$
          $= 1$ M $= 495$ kNm
          $= 2$ M $= 94$ kNm
          $= 3$ M $= 133 \cdot 5$ kNm
          $= 4$ M $= 168$ kNm

Similarly for BC

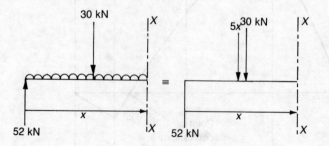

$$SF_{XX} = 52 - 30 - 5x$$

> when $x = 4$ SF = 2 kN
> when $x = 10$ SF = -28 kN

$$M_{XX} = 52x - \frac{5x^2}{2} - 30(x - 4)$$

> when $x = 4$ M = 168 kNm
> $x = 6$ M = 162 kNm
> $x = 8$ M = 128 kNm
> $x = 10$ M = 90 kNm

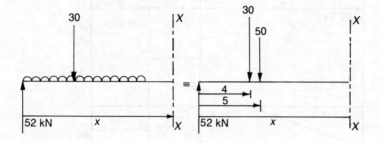

For CD

$$SF_{XX} = 52 - 50 - 30 = -28 \text{ kN}$$
$$\text{and } M_{XX} = 52x - 30(x - 4) - 50(x - 5)$$

> when $x = 10$,   M = 90 kNm
> when $x = 14$,   M = -22 kNm

Finally from D to E

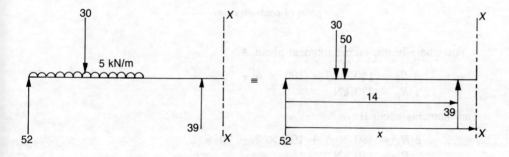

$$SF_{XX} = 52 - 50 - 30 + 39 = +11 \text{ kN}$$
$$\text{and } M_{XX} = 52x - 30(x - 4) - 50(x - 5) + 39(x - 14)$$

> when $x = 14$,   M = -22 kNm
> when $x = 16$,   M = 0

Example 2.25

A beam ABCD is simply supported at A and C which are 8 m apart. A load of 30 kN/m is uniformly distributed over the portion AB which is 6 m long. Concentrated loads of 100 kN and 70 kN act at B and D respectively where D is on an overhang 2 m from C. Sketch the shear force and bending moment diagrams and calculate the value and position of the maximum bending moment and the point of contraflexure.

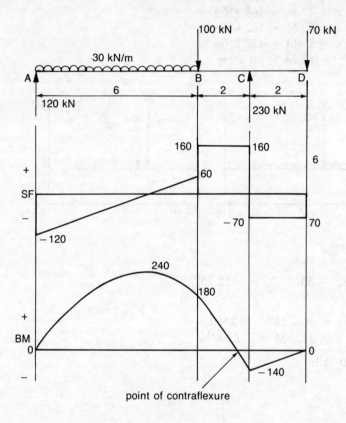

For whole beam, taking moment about A

$$8 R_B = 180 \times 3 + 100 \times 6 + 70$$
$$\therefore R_B = 230 \text{ kN}$$

and moments about B

$$8 R_A = 180 \times 5 + 100 \times 2 - 70 \times 2$$
$$\therefore R_A = 120 \text{ kN}$$

Check $R_A + R_B = 120 + 230 = 350$ kN

For AB

$$SF_{XX} = -120 + 30x$$

$$BM_{XX} = 120x - \frac{30x^2}{2}$$

For BC

$$SF_{XX} = -120 + 180 + 100$$
$$= +160 \text{ kN}$$
$$BM_{XX} = 120x - 180(x - 3) - 100(x - 6)$$

For CD

$$SF_{XX} = -120 - 230 + 180 + 100$$
$$= -70 \text{ kN}$$
$$BM_{XX} = 120x - 180(x - 3) - 100(x - 6) + 230(x - 8)$$

From the diagram the maximum bending moment occurs on AB when $M = 120x - 15x^2$.

Bending moment is a maximum when $\dfrac{dM}{dx} = 0$

i.e. when $120 - 30x = 0$ or when $x = 4$ m

when $x = 4$ $M_{max} = 120 \times 4 - 15 \times 4^2 = \underline{240 \text{ kNm}}$

The point of contraflexure occurs when

$$M = 0 \quad \text{or when } 120x - 180(x - 3) - 100(x - 6) = 0$$

that is when $x = 7\frac{1}{8}$ m

Example 2.26   Guided Solution

A beam ABCDE is simply supported at B and D, and carries point loads of 100 kN and 30 kN at C and E respectively, together with uniformly distributed loads of 40 kN/m over the portion A to B and 10 kN/m over the portion C to E. AB = 1 m, BC = 4 m, CD = 2 m and DE = 3 m.

   Draw the shear force and bending moment diagrams and determine the magnitude and position of

a the maximum shear force
b the maximum bending moment

Determine also the positions of the points of contraflexure.

Step 1. Draw a diagram of the beam, and construct axes for SF and BM directly beneath.

Step 2. Calculate the value of the reactions at B and D

$$(R_B = 57 \cdot 5 \text{ kN}, \quad R_D = 162 \cdot 5 \text{ kN})$$

Step 3. Define the sign conventions to be used. Say $\lrcorner$ +ve SF $\bigcirc$ +ve BM

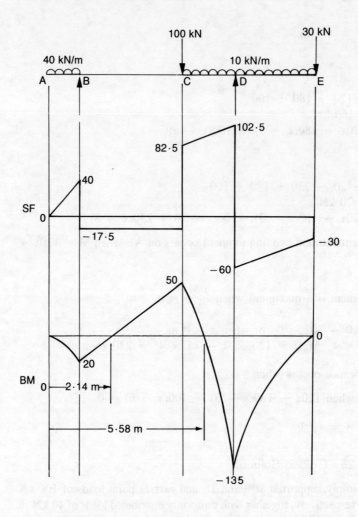

Step 4. Consider the portion of the beam from A to B, and write down expressions for SF and BM at a section XX a distance x from A and contained within the portion AB

$$(SF_{XX} = 40x, \quad BM_{XX} = -20x^2)$$

Step 5. Substitute values of x applicable to the portion AB into each expression obtained in step 4 so that the SF and BM diagram can be drawn for this portion on the prepared axes.

Step 6. Repeat steps 4 and 5 for each portion of the beam B to C, C to D and D to E.

$$(B \text{ to } C \quad SF_{XX} = 40 - 57\cdot5 = -17\cdot5 \text{ kN}$$
$$BM_{XX} = -40(x - \tfrac{1}{2}) + 57\cdot5(x - 1)$$
$$= 17\cdot5(x - 37\cdot5)$$

$$C \text{ to } D \quad SF_{XX} = 40 + 100 - 57\cdot5 + 10(x - 5)$$

$$BM_{XX} = -40(x - \tfrac{1}{2}) + 57 \cdot 5(x - 1) - 100(x - 5)$$
$$- 10\,\frac{(x - 5)^2}{2}$$

D to E $SF_{XX} = 40 - 57 \cdot 5 + 100 - 162 \cdot 5 + 10(x - 5)$

$$BM_{XX} = -40(x - \tfrac{1}{2}) + 57 \cdot 5\,(x - 1) - 100\,(x - 5)$$
$$+ 162 \cdot 5(x - 7) - 10\,\frac{(x - 5)^2}{2}$$

Step 7. As a check to your calculation remember that

a the SF diagram changes abruptly at a point load by the amount of the point load, and gradually over a UD load by the amount of the UD load

b the BM diagram reaches peak values (+ and −) when the SF diagram crosses the zero axis

c the BM is always zero at the ends of a beam except for the built-in end of a cantilever.

Step 8. Ascertain from your diagrams the points of contraflexion (i.e. points where the BM = 0). Determine the values of x at which they occur by equating the BM expression *for the correct portions of the beam* to zero. (2.14, 5.58).

Your finished diagrams should look like the diagram on page 78.

From the diagrams Maximum SF $102 \cdot 5$ kN   when $x = 7$ m.

Maximum BM   135 kNm   when $x = 7$ m.

Problems 2.9

1 Draw the shear force and bending moment diagrams for the beams shown in question 1 and 2 of problems 2.7.

2 For the beam shown sketch the shear force and bending moment diagrams and calculate the position and magnitude of the maximum bending moment.

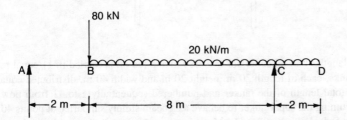

3 A beam ABC is simply supported at A and B and carries a uniformly distributed load of 25 kN/m over the portion AB which is 8 m long. At C which is on an overhang 2 m from B there is a concentrated load of 45 kN. Determine the position and magnitude of the maximum bending moment and the position of the point of contraflexure.

4 A beam l m long is supported symmetrically with equal overhangs at each end of 2 m. Each overhang carries a concentrated load of W at its extreme end, and there is a concentrated load of W in the centre of the beam. If the bending moment at mid-span is equal to that at each support calculate the distance between the supports. Sketch the SF and BM diagrams and find the position of the points of contraflexure.

5 A yacht is lifted from a dock using two slings. Assuming the yacht has uniform weight per unit length, calculate the best positions for the slings if the bending moment due to its own weight is to be kept as small as possible. For this arrangement draw the SF and BM diagrams.

6 The jib of a crane may be regarded as a cantilever of length 16 m. Calculate the maximum load that can be lifted at the extreme end of the jib, when it is horizontal, if the weight of the jib is 100 kN per metre length and
   a the maximum SF must not exceed 2500 kN
   b the maximum BM must not exceed 29 MNm

7 A steel shaft ABCDE, 5 m long and weighing 1 kN/m carries three pulleys at B, C and E as shown. The shaft is supported in bearings at A and D which are $4\cdot5$ m apart. Pulley B weighs $1\cdot5$ kN and experiences a total belt tension of 1 kN vertically down, pulley C weighs 1 kN and experiences a total belt tension of $0\cdot5$ kN vertically down and pulley E weighs 2 kN and experiences a total belt tension of $0\cdot9$ kN vertically upwards. Draw the SF and BM diagrams for the shaft and calculate the value and position of the maximum bending moment.

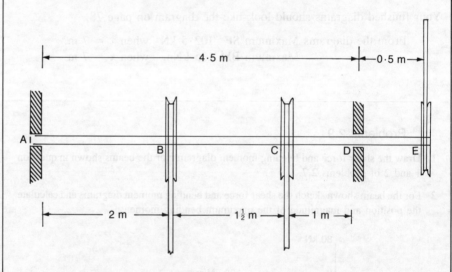

8 The total length of an oil supertanker is 220 m and the oil is carried in 11 equal rectangular tanks, each of length 20 m, height 20 m and width 40 m, distributed equally along the total length of the tanker and numbered sequentially 1 to 11 from bow to stern. Assuming the supertanker to behave as a beam simply supported at points 40 m from each end, draw the SF and BM diagrams and calculate the maximum bending moment and the point at which it acts when tanks 4, 5 and 6 are completely empty and the remainder completely full of oil. Take specific gravity of oil as $0\cdot8$. Ignore the weight of the tanker itself.

2.10 Torsion

When a pure torque is applied to a round shaft one end of the shaft twists relative to the other. Each cross-sectional element of the shaft is in a state of pure shear, since each element is tending to rotate relative to the next. The shearing stresses thereby induced in the shaft produce a moment of resistance, equal and opposite to the applied torque.

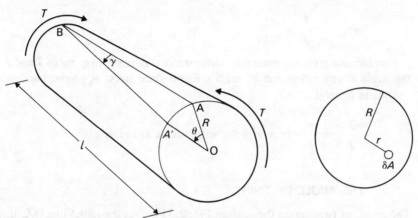

Fig. 2.33

The theory of twisting makes the following assumptions:

a The material obeys Hooke's law, i.e. the shear stress at any point is proportional to the shear strain at that point.

b Radial lines remain straight after twisting.

c The shaft is of circular solid or hollow section.

STRESSES DUE TO TWISTING

From the first assumption it follows that the strain (and hence the stress) is directly proportional to the radius. Consider the shaft in Fig. 2.33. A pure torque T is applied to a shaft of radius R and length l producing an angle of twist θ. (A line AB marked on the axial length of the shaft becomes A$'$B after twisting.) If the shear stress at the surface of the shaft is τ_{max} then the shear stress at some radius r will be $\tau_{max} \times r/R$.

The force acting on an elemental area δA at radius $r = \dfrac{r}{R} \times \tau_{max} \times \delta A$

The moment of this force about O $= \dfrac{r}{R} \times \tau_{max} \times \delta A \times r$

Total moment of resistance $= \displaystyle\int r^2 \dfrac{\tau_{max}}{R} \, dA = \dfrac{\tau_{max}}{R} \int r^2 \, dA$

Now $\int r^2 dA$ is the polar second moment of area of the cross section and is given the symbol J (see Appendix 3). For a solid circular cross-section $J = \pi D^4/32$ where D is the diameter of the shaft.

Therefore total moment of resistance $= \dfrac{\tau_{max} \times J}{R}$

But, in the equilibrium state, total moment of resistance = applied torque T

Therefore $T = \dfrac{\tau_{max} \times J}{R}$

or $\qquad \dfrac{T}{J} = \dfrac{\tau_{max}}{R}$ $\hfill$ (2.22)

This formula gives the maximum shear stress at radius R in terms of T and J. Clearly the stress at any radius can be readily found since stress is proportional to radius. Thus in general

$$\dfrac{T}{J} = \dfrac{\tau}{r} \text{ where } \tau \text{ is the shear stress at radius } r \hfill (2.23)$$

THE ANGLE OF TWIST

The action of twisting on the shaft in Fig. 2.33 moves the radial line OA, through an angle of twist to a new position OA$'$. At the same time the axial line AB moves through an angle γ to a new position A$'$B.

The shear strain experienced by the shaft $= \dfrac{AA'}{AB} = \dfrac{R\theta}{l} = \gamma_{max}$ $\hfill$ (2.24)

where γ_{max} is the maximum strain at radius R measured in radians.

Thus the value of γ at any radius r is given by:

$$\gamma = \dfrac{r\theta}{l} \hfill (2.25)$$

The ratio $\dfrac{\text{shear stress, } \tau}{\text{shear strain, } \gamma} = G$, the modulus of ridigity or $\gamma = \tau/G$ $\hfill$ (2.26)

Equating (2.25) and (2.26)

$$\dfrac{r\theta}{l} = \dfrac{\tau}{G} \hfill (2.27)$$

This formula gives the shear stress at any radius in terms of G and the angle of twist per unit length measured in radians.

Combining equations (2.23) and (2.27) gives a general equation for the torsion of shafts of circular cross-section.

$$\dfrac{T}{J} = \dfrac{\tau}{r} = \dfrac{G\theta}{l} \hfill (2.28)$$

where T = applied torque in Nm
$\quad\quad J$ = polar second moment of area in m^4
$\quad\quad \tau$ = shear stress at radius r in N/m^2
$\quad\quad r$ = radius at which shear stress is required
$\quad\quad G$ = modulus of rigidity in N/m^2
$\quad\quad \theta$ = angle of twist in radians
$\quad\quad l$ = length of shaft at which θ is measured.

Example 2.27

A hollow steel shaft is 225 mm outside diameter and 150 mm inside diameter. Calculate:

a The maximum power this shaft can transmit at 150 rev/min if the max shear stress is not to exceed 70 MN/m^2.

b The diameter of a solid shaft of the same material which would transmit the same maximum power at the same speed with the same stress.

a J for a hollow shaft $= \dfrac{\pi(D^4 - d^4)}{32} = \dfrac{\pi(0\cdot225^4 - 0\cdot150^4)}{32}$

$$= 2\cdot02 \times 10^{-4} \text{ m}^4$$

Using $\dfrac{T}{J} = \dfrac{\tau}{r}$, $T_{max} = \dfrac{\tau_{max} \times J}{D/2} = \dfrac{70 \times 10^6 \times 2\cdot02 \times 10^{-4}}{0\cdot225/2}$

$$= 1\cdot257 \times 10^5 \text{ Nm}$$

Max. power $= T_{max}\, w = T_{max} \times \dfrac{2\pi N}{60} = \dfrac{1\cdot257 \times 10^5 \times 2\pi \times 150}{60}$

$$= 1974\cdot5 \text{ kW}$$

b For a solid shaft $J = \dfrac{\pi D^4}{32}$

$\quad\quad\quad\quad$ and $J = \dfrac{Tr}{\tau}$

$\quad$ therefore $\dfrac{\pi D^4}{32} = \dfrac{T_{max} \times D/2}{\tau_{max}}$

$\quad\quad$ hence $D^3 = \dfrac{32\,T_{max}}{2\pi\tau_{max}} = \dfrac{32 \times 1\cdot257 \times 10^5}{2\pi \times 70 \times 10^6} = 9\cdot15 \times 10^{-3}$

$\quad\quad$ therefore $D = 0\cdot209$ m
$\quad\quad\quad\quad$ or $D = 209$ mm

Example 2.28

A 150 mm diameter shaft runs at 120 rev/min and transmits 400 kW. Calculate the maximum shear stress of the material. Find also the angle of twist, in degrees, in a length of 3 m. $G = 82 \cdot 5$ kN/mm^2.

$$\text{Power} = T\omega$$

$$\text{or } P = \frac{T \times 2\pi N}{60}$$

$$\text{therefore } T = \frac{60P}{2\pi N} = \frac{60 \times 400 \times 10^3}{2\pi \, 120}$$

$$= \frac{10^5}{\pi}$$

$$\text{Using } \frac{T}{J} = \frac{\tau}{r}, \tau = \frac{Tr}{J} = \frac{10^5 \times 0 \cdot 075}{\pi \times 4 \cdot 97 \times 10^{-5}} = 48 \text{ MN/m}^2$$

$$J = \frac{\pi D^4}{32} = \frac{\pi (0 \cdot 15)^4}{32} = 4 \cdot 97 \times 10^{-5} \text{ m}^4$$

$$\text{Using } \frac{\tau}{r} = \frac{G\theta}{l}, \theta = \frac{\tau l}{Gr} = \frac{48 \times 10^6 \times 3}{82 \cdot 5 \times 10^9 \times 0 \cdot 075} = 0 \cdot 023 \, 27 \text{ radians}$$

$$= 1° \, 20'$$

Problems 2.10

1 A mild steel shaft, 50 mm diameter and $0 \cdot 6$ m long, rigidly fixed at one end, is subjected to a torque of 135 Nm applied in the plane of the cross-section at the other end. Calculate:
    a the maximum shear stress
    b the angle of twist
    c the shear stress at a radius of 18 mm.
    $G = 80$ kN/mm^2

2 A hollow shaft, 75 mm external diameter and 50 mm internal diameter, transmits 15 kW at 1000 rev/min. Calculate the maximum shear stress set up in the shaft.

3 **a** Determine the diameter of a solid steel shaft which can transmit 45 kW at 400 rev/min if the maximum shear stress is not to exceed 45 N/mm^2.
    b If the weight of the shaft is reduced by 50% by drilling an axial hole through the centre, what will be the maximum stress if the power and the speed remain unchanged?

4 A solid shaft is to transmit 750 kW at 200 rev/min. If the shaft is not to twist more than 1° on a length of twelve diameters, and the shear stress is not to exceed 45 N/mm^2, calculate the minimum shaft diameter required. $G = 80$ GN/m^2
    (*Hint*: There are two independent conditions to be satisfied in this problem. The torque is limited by *both* the twist *and* the shear stress — calculate for both.)

5 A rule often used in shaft design states that 'the angle of twist should not exceed 1° on a length equal to 20 diameters'. What stress in the material does this imply if the modulus of rigidity is equal to 80 kN/mm^2?

6 a Determine the maximum allowable power which can be transmitted by a 150 mm diameter shaft running at 240 rev/min when the permissible shear stress is 50 N/mm^2.

 b The shaft shown has a coupling on it which has 6 bolts on a 260 mm diameter pitch circle.

 Determine the diameter of the bolts if the maximum shear stress in the bolts must not exceed 100 N/mm^2, and the power transmitted is 314 kW at 500 rev/min.

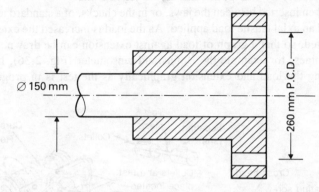

Fig. 2.34

7 Determine the external diameter of a hollow shaft to transmit 7·5 MW at 180 rev/min, the maximum permissible shear stress being 45 N/mm^2 and the internal diameter being 0·6 of the external.

8 The propeller shaft of an aircraft engine is steel tubing of 75 mm external and 60 mm internal diameter. The shaft is to transmit 150 kW at 1650 rev/min. The failing stress in shear for this shaft is 135 N/mm^2. What is the factor of safety?

2.11 Tensile testing

A tensile test is one of the most common methods of determining whether a particular material is suitable for a certain application. The test consists of placing a sample of the material between the jaws of a machine capable of applying sufficient direct tensile load to extend the sample eventually to its breaking point. The sample of material must conform to measurements laid down by British Standards (BS 18 parts 1−4 for metals) so that the results from the testing of different materials can be easily compared. Fig. 2.35 shows a typical test sample circular cross-section that can easily be produced by turning on a lathe.

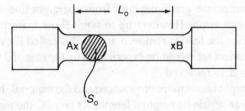

Fig. 2.35 Note: L_0 = original gauge length, S_0 = original cross-sectional area of gauge length

A strict relationship is maintained between S_o and L_o. So that $L_o = 6 \cdot 65 \sqrt{S_o}$ or L_o approx. proportional to $5 \times$ diameter. Recommended diameters are tested in the British Standard.

Other shapes of test sample are possible and have standardized dimensions, for example, test specimens from flat bar and cut from the side of a circular tube.

To prepare test specimens of circular cross-section it is standard practice to mark L_o on the sample, usually by fine punch marks, or scribed lines.

The specimen is then inserted between the jaws, or in the chucks, of a standard tensile-testing machine and an axial tensile load applied. As the load is increased the extension of the sample is noted, so that a graph of load against extension can be drawn. Some tensile-testing machines, for example the Hounsfield tensometer (Fig. 2.36), have a facility for recording the load and extension graphically as the test is in progress.

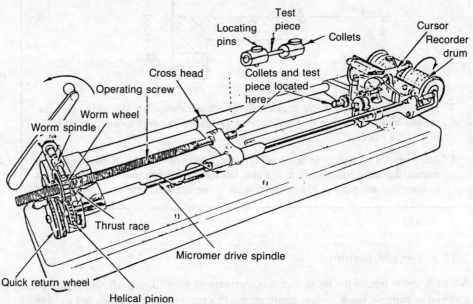

Fig. 2.36 The Hounsfield 'W' type tensometer

A typical load/extension graph obtained from such a machine for a sample of ductile material such as mild steel is shown in Fig. 2.37.

The graph can be divided into a number of distinct portions as follows:

O to A A direct proportionality exists between the load and extension as indicated by the straight line. Point A is the *limit of proportionality*, and Hooke's law states that load is proportional to extension (or stress is proportional to strain) up to the limit of proportionality.

A to B Over this very short portion the graph deviates from a straight line so that load is no longer proportional to extension. However up to point B the material will still return to its original length when the load is removed and so B is called the *elastic limit*. Beyond this point some permanent set would be experienced by the material which would remain even though the load is removed.

B to C The deviation from a straight line is more pronounced and the material, having exceeded the elastic limit, will never regain its original length. At point C the material

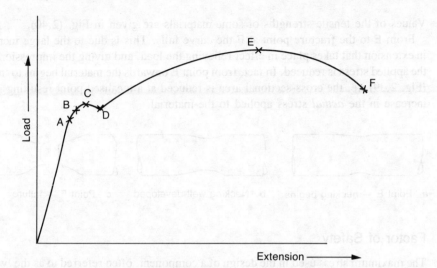

Fig. 2.37

is observed to suddenly 'give', in that a large increase in extension takes place with
little or no increase in load. In fact in some cases the load may appear to fall after
point C which is called the *Yield Point*.

The stress at the yield is called the *Yield Stress*.

In some materials there may be more than one yield point. Fig. 2.38 shows the
stress/strain graph for a material for which there is an upper and lower yield point.

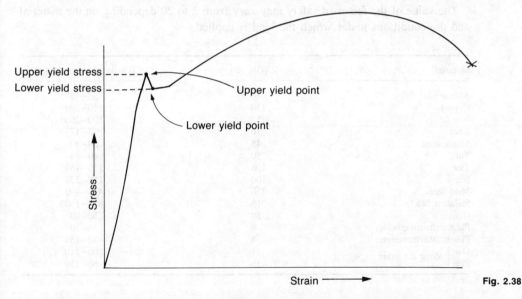

Fig. 2.38

Referring again to Fig. 2.37: as the material is further stressed point E is reached
which is the maximum value of load that the material can withstand. This value is termed
'the Ultimate Tensile Stress' (UTS) or '*Tensile Strength*', where

$$\text{tensile strength} = \frac{\text{maximum load}}{\text{original cross-sectional area}}.$$

Values of the tensile strengths of some materials are given in Fig. (2.40).

From E to the fracture point at F the curve falls. This is due to the large increase in extension that takes place in effect relieving the load, and giving the impression that the applied stress is reduced. In fact, from point E onwards the material begins to 'neck' (Fig. 2.39) i.e. the cross-sectional area is reduced at a localised point resulting in an increase in the *actual* stress applied to the material.

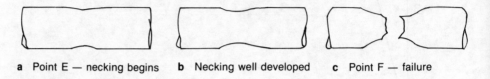

Fig. 2.39

a Point E — necking begins   **b** Necking well developed   **c** Point F — failure

Factor of Safety

The maximum stress used in the design of a component, often referred to as the 'working stress', is considerably less than the ultimate stress to allow for possible faults in the material, overloading, stress concentrations, wear, or just oversimplification of the design. The ratio ultimate stress divided by working stress is called the 'Factor of Safety'.

$$\text{i.e. Factor of Safety} = \frac{\text{UTS}}{\text{working stress}}$$

The value of the factor of safety may vary from 2 to 20 depending on the material and the conditions under which the load is applied.

Material	E (GN/m²)	Tensile strength (MN/m²)
Aluminium	70.5	90−100
Copper	130	120−200
Iron (cast)	152	100−200
Lead	16	12−17
Magnesium	45	60−80
Tin	50	20−35
Zinc	108	110−150
Brass	100	150−270
Mild Steel	210	430−490
Stainless Steel	215	800−1500
Glass	80	30−90
Plastic (thermoplastic)	6	30−70
Plastic (thermosetting)	4	40−150
Oak ⎰ along the grain	11	60−110
Pine ⎱	16	60−110

Fig. 2.40

PROOF STRESS

In some materials, notably alloy steels and non-ferrous metals it is difficult to identify the exact position of the limit of proportionality since there is such a gradual change of slope and virtually no yield point. In such cases an artificial measure of the elastic limit stress called the 'proof stress' is defined as the value of stress giving a permanent

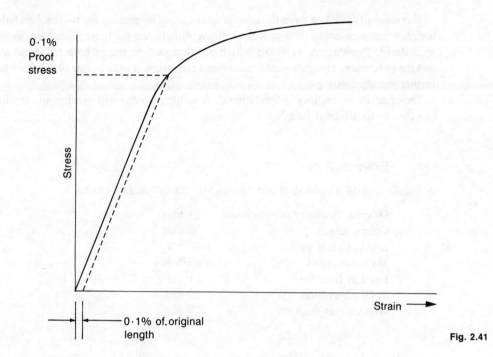

0·1%
Proof
stress

Stress

Strain

0·1% of.original
length

Fig. 2.41

set (or strain), usually 0·1% (or 0·001). Fig. 2.41 clearly shows how the proof stress can be calculated from a stress-strain curve.

As mentioned previously, the specimen may begin to neck at the maximum load if it is a ductile material, and this causes the specimen to extend with apparently decreasing load. In fact we have seen that the *actual* stress increases due to the rapidly reducing area of cross-section.

A very important property of a material that can be measured in a tensile test is that of 'ductility'. A material is said to be ductile if it can be drawn out without fracturing and is specified by the increase in gauge length, *lo*, expressed as a percentage;

$$\text{i.e. Percentage elongation} = \frac{\text{increase in length}}{\text{original length}} \times 100$$

$$= \frac{l - lo}{lo} \times 100$$

and by the percentage reduction of area at the fracture;

$$\text{i.e. Percentage reduction in area} = \frac{(\text{original area} - \text{area at fracture})}{\text{original area}} \times 100$$

$$= \frac{So - S}{So} \times 100$$

where l = extended length after fracture and S = cross-sectional area at the fracture point.

The value of l is taken from the sample after testing by placing the two broken halves together and measuring their new overall length between the marks that were used to indicate lo. Devices are available which are designed to accept both halves in a jig and the percentage elongation and percentage reduction in area can be read off without further calculation.

The opposite of ductility is 'brittleness'. A brittle material will break with very little increase in its original length.

Example 2.29

A tensile test on a mild steel specimen gave the following results:

Original diameter of specimen	14 mm
Gauge length	50 mm
Load at yield point	46 kN
Maximum load	69·5 kN
Load at fracture	51 kN
Total elongation	12 mm
Diameter at fracture	9 mm

Find

   a the stress at the yield point
   b the maximum stress
   c the percentage elongation
   d the percentage reduction in area
   e the nominal and actual stresses at fracture.

a Stress at Yield Point
$$= \frac{\text{load at yield point}}{\text{original c.s.a.}}$$

$$= \frac{46 \times 4}{\pi \times (0 \cdot 014)^2} = 299 \text{ MN/m}^2$$

b Maximum stress
$$= \frac{\text{maximum load}}{\text{original c.s.a.}}$$

$$= \frac{69 \cdot 5 \times 4}{\pi (0 \cdot 014)^2} = 450 \text{ MN/m}^2$$

c Percentage elongation
$$= \frac{12}{50} \times 100 = 24\%$$

d Percentage reduction in area $= \dfrac{\pi/4 (0 \cdot 0142^2 - 0 \cdot 009^2)}{\pi/4 \times (0 \cdot 014)^2} \times 100$

$$= 58 \cdot 7\%$$

e Nominal stress at fracture $= \dfrac{\text{load at fracture}}{\text{original c.s.a.}}$

$$= \frac{51 \times 4}{\pi \, 0 \cdot 014^2} \qquad = \underline{331 \text{ MN/m}^2}$$

Actual stress at fracture $= \dfrac{\text{load at fracture}}{\text{area of fracture}}$

$$= \frac{51 \times 4}{\pi \, (0 \cdot 009)^2} \qquad = \underline{800 \text{ MN/m}^2}$$

2.12   Hardness testing

The hardness of a material may be described as the surface resistance of the material to indentation. Several types of test exist but each entails causing an indentation in the surface of the material which results in a hardness number based on the size of the indentation caused by a given load, or the size of the load required to cause a given indentation.

The *Brinell Test* uses a hardened chrome-steel ball which is pressed into the surface by a load applied with a purpose-built machine. The diameter of the indentation is then accurately measured using a travelling microscope and a Brinell Hardness number (HB) for the material calculated, where:

$$\text{HB} = \frac{\text{applied load in kg}}{\text{spherical area of impression}}$$

The control of Brinell Hardness tests should be in accordance with British Standard BS 240. This lists the conditions that must apply if accurate results are to be achieved.

In order that the hardness numbers of different materials can be compared different ball diameters with corresponding loads are used to give an HB number that makes comparisons more meaningful. For example, mild steel would require a ball diameter, D, of 10 mm with an applied load, P, of 3000 kg to give a P/D^2 of 30, whereas pure lead would require a 100 kg load with the same size ball, or a 25 kg load with a 5 mm diameter ball, to give a P/D^2 of $1 \cdot 0$.

Each material, then, requires a P/D^2 ratio to be chosen that corresponds to its approximate hardness number. Fig. 2.42 indicates the desirable value according to BS 240.

When specifying the Brinell Hardness of a material it is important to quote the diameter of the ball used and the load that was applied. For example 118 HB 10/1000 indicates a Brinell Hardness number of 118 achieved by using a 10 mm diameter ball with an applied load of 1000 kg.

This difficulty of a variable scale of hardness together with problems that occur when using a ball as the indenter are largely overcome in the *Vickers Hardness Test*. In this test an inverted diamond pyramid is used with an angle between opposing faces of 136° (Fig. 2.43a).

Approximate Hardness Number	P/D^2
Above 160	30
160–60	10
60–20	5
Up to 20	1

Fig. 2.42

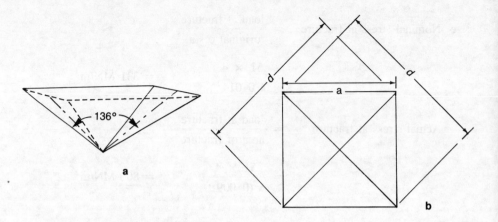

Fig. 2.43

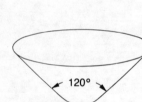

Fig. 2.44

The load is applied automatically for 30−35 seconds by a system of levers and can range from 1−100 kgf, however the hardness numbers do not depend on the load which is governed by the hardness of the material. Thus some machines are capable of only applying one load, and it is recommended that the average of several indentations are then made. The indentation produced is measured across each diagonal (Fig. 2.43b) and the mean d ascertained. From this dimension the side a of the square base and the slant length ℓ can be calculated. The Vickers Hardness (HV) number can then be calculated from the formula, $\dfrac{\text{HV} = 2\,\ell\,\sin\theta}{d^2}$, which represents the surface area of the indentation.

Again it is common practice to quote the Vickers Hardness with the load applied to produce it. Thus 640 HV 30 indicates a Vickers Hardness of 640 with an applied load of 30 kgf.

The *Rockwell Hardness Test* uses either a ball (T) or a diamond cone (N) with a vertex angle of 120° (Fig. 2.44). The relevant British Standard is BS 891.

Most machines using the Rockwell principle are graduated in a number of scales depending on the combination of indenter and applied load. The most common scales are usually denoted C and B, with C the usual one for the harder metals. The Rockwell Hardness number is obtained by measuring the depth of penetration of the indenter after the application of a combination of loads. A preliminary load F_0 is first applied (Fig. 2.45a) and when an equilibrium situation has been achieved an indicator on the measuring dial is set to a prescribed datum marked 'set'. An additional load, F is now applied (Fig. 2.45b) which results in an increased penetration. Again when the

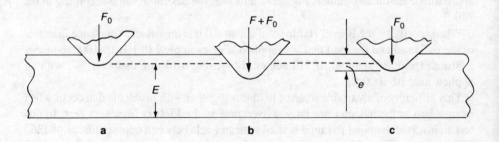

Fig. 2.45

equilibrium state has been reached, the additional load is removed but the preliminary load F_0 remains (Fig. 2.45c). The material exhibits a recovery to some extent, and the permanent increase in the depth of penetration, e, is used to calculate the Rockwell Hardness (HR).

So that $HR = E - e$ where E is a constant $= 100$ units.

As in the case of the other types of test it is common practice to quote the Rockwell Hardness with an indication of the conditions applicable.

Thus 70 HRC 30 N indicates Rockwell Hardness 70 measured on C scale with 30 kgf load and diamond-cone indenter. The Brinell, Vickers and Rockwell Scales of hardness can be compared using tables notably those prepared by British Standards Institution in BS 860.

2.13   Impact testing

Impact tests are designed to measure toughness or a material's ability to withstand shock loading and most tests measure the energy required to fracture a standard test piece with a sharp blow. It is important that the sample is broken completely and not just bent, and so to ensure that this happens a notch is cut into the sample. British Standard BS 131 details the procedures and conditions to observe in all Notched Bar Tests.

In the Izod Test the sample may be of square or circular section, and there may be one, two or three notches as required.

Fig. 2.46 shows a standard square-section bar with single notch.

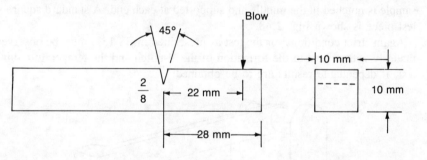

Fig. 2.46

The sample of material to be tested is prepared and mounted in a firm base with the axis of the notch perpendicular to the direction of the blow. A pendulum (Fig. 2.47) of given mass M at a specified height, h_1 and thus with certain potential energy, mgh_1 swings down and strikes the test piece at a predetermined position, snapping it cleanly. The pendulum then continues its swing, coming to rest at a height h_2 which is less than h_1. There has thus been a loss of energy of $mg(h_1 - h_2)$ which is assumed to be the energy absorbed in fracturing the test piece. In practice the pendulum can be made to activate a loose pointer that records the energy required to fracture the sample on a specially graduated scale. Normally the test should be repeated and an average value taken, which is why test pieces are available with up to three notches, cut into different sides of the samples.

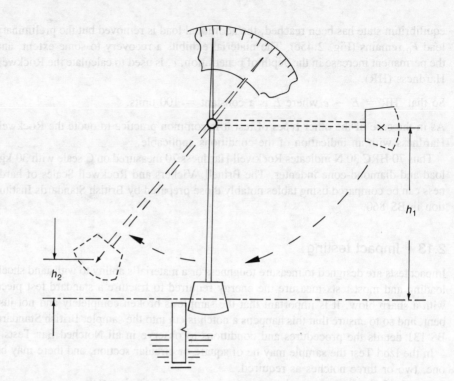

Fig. 2.47

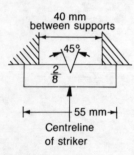

40 mm
between supports

45°

$\frac{2}{8}$

55 mm

Centreline
of striker

Fig. 2.48

The *Charpy V-notch Impact Test* uses the same principle as the Izod Test, but the sample is notched in the middle and supported at each end. A standard square-section test piece is shown Fig. 2.48.

Again strict conditions for the test as prescribed in BS 131, must be observed, particularly with regard to the formation of the V-notch and the temperature during the test, if dependable results are to be obtained.

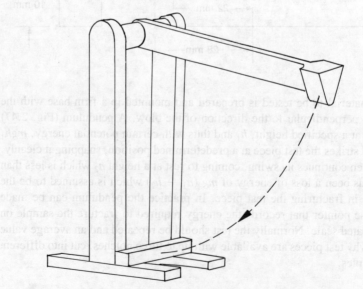

Fig. 2.49

3 Dynamics

The contents of this chapter enable you to solve problems involving angular motion, simple harmonic motion, and linear and angular kinetic energy.

3.1   Angular motion

THE RADIAN

One radian is the angle subtended at the centre of a circle by an arc whose length is equal to the radius of the circle. Thus an arc length of r subtends 1 radian, therefore an arc length of $2\pi r$ (1 revolution) subtends 2π radians, i.e.

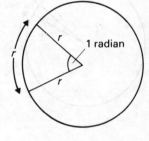

$$2\pi \text{ radians} = 360°$$

$$\text{or 1 radian} = \frac{360°}{2\pi}$$

$$= \frac{360}{2 \times 3 \cdot 142}$$

$$= 57° \ 18'$$

ANGULAR QUANTITIES

Angular displacement is then denoted by θ and is measured in radians.

Angular velocity is the rate of change of angular displacement, $d\theta/dt$ is denoted by ω and measured in radians per second (rad/s).

Angular acceleration is the rate of change of angular velocity,

$$\frac{d\omega}{dt} \left(= \frac{d^2\theta}{dt^2} \right)$$

is denoted by α and measured in radians per second per second (rad/s^2).

Also $\dfrac{d\omega}{dt} = \dfrac{d\theta}{dt} \times \dfrac{d\omega}{d\theta} = \omega \dfrac{d\omega}{d\theta}$

therefore angular acceleration $= \omega \dfrac{d\omega}{d\theta}$

The equations of angular motion for *constant* angular acceleration are similar to those used for linear motion with constant acceleration.

The table below gives a comparison of the two sets of equations.

Linear motion	Angular motion	
$v = u + at$	$\omega_2 = \omega_1 + at$	(3.1)
$s = \dfrac{u + v}{2} \times t$	$\theta = \dfrac{\omega_1 + \omega_2}{2} \times t$	(3.2)
$s = ut + \frac{1}{2}at^2$	$\theta = \omega t + \frac{1}{2}\alpha t^2$	(3.3)
$v^2 = u^2 + 2as$	$\omega_2^2 = \omega_1^2 + 2\alpha\theta$	(3.4)

RELATIONSHIP BETWEEN LINEAR AND ANGULAR QUANTITIES

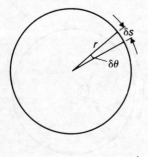

Given that a small angular displacement $\delta\theta$ gives rise to a small displacement of δs on an arc of radius r, then

$$\delta s = r\delta\theta$$

If the change takes place in an increment of time δt, then

$$\frac{\delta s}{\delta t} = r\frac{\delta\theta}{\delta t}$$

In the limit as δt tends to zero

$$\frac{ds}{dt} = r\frac{d\theta}{dt}$$

but $\dfrac{ds}{dt} = v$ and $\dfrac{d\theta}{dt} = \omega$ therefore $v = r\omega$ m/s   (3.5)

or $\quad \omega = \dfrac{v}{r}$ rad/s   (3.6)

Differentiating equation (3.5) with respect to t

$$\frac{dv}{dt} = \frac{d}{dt}(r\omega)$$

$$= r\frac{d\omega}{dt}$$

but $\dfrac{dv}{dt} = a$ (linear acceleration) and $\dfrac{d\omega}{dt} = \alpha$ (angular acceleration)

therefore $a = r\alpha$ (3.7)

$$\text{or} \quad \alpha = \frac{a}{r} \tag{3.8}$$

Thus any linear quantity can be changed into its angular equivalent by simply dividing by the radius, i.e.

$$\theta = \frac{s}{r}, \quad \omega = \frac{v}{r}, \quad \alpha = \frac{a}{r}$$

Example 3.1

A car has wheels of $0 \cdot 4$ m dia. and is uniformly accelerated in a straight line from 50 km/h to 80 km/h in 10 seconds. Calculate:

a the angular acceleration of the wheels

b the number of revolutions made by the wheels during acceleration.

$$u = 50 \text{ km/h} = \frac{50 \times 10^3}{60 \times 60} \text{ m/s} = 13 \cdot 89 \text{ m/s}$$

$$v = 80 \text{ km/h} = \frac{80 \times 10^3}{60 \times 60} \text{ m/s} = 22 \cdot 22 \text{ m/s}$$

If the car has a velocity of v m/s then any point on the rim of a wheel has a tangential velocity of v m/s (for example consider the point of contact of the wheel with the ground). If ω is the angular velocity of the wheel, then

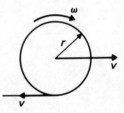

$$\omega = \frac{v}{r}$$

$$\text{therefore} \quad \omega_1 = \frac{13 \cdot 89}{0 \cdot 2} = 69 \cdot 45 \text{ rad/s}$$

$$\text{and} \quad \omega_2 = \frac{22 \cdot 22}{0 \cdot 2} = 111 \cdot 1 \text{ rad/s}$$

Using $\omega_2 = \omega_1 + \alpha t$

$$\alpha = \frac{\omega_2 - \omega_1}{t} = \frac{111 \cdot 1 - 69 \cdot 45}{10} = 4 \cdot 17 \text{ rad/s}^2$$

Using $\theta = \dfrac{\omega_1 + \omega_2}{2} \times t = \dfrac{69 \cdot 45 + 111 \cdot 1}{2} \times 10 = 902 \cdot 75$ radians

There are 2π radians in 1 revolution

therefore number of revolutions made by wheel $= \dfrac{902 \cdot 75}{2\pi} = 143 \cdot 7$

Problems 3.1

1 A flywheel increases its speed from 30 to 60 rev/min in 10 s. The diameter of the wheel is 3500 mm. Calculate:
  a the angular acceleration
  b the number of revolutions made during the 10 seconds
  c the linear acceleration of a point on the rim of the flywheel.

2 A vehicle with wheels 900 mm diameter has its speed reduced from 50 km/h to 30 km/h while it travels a distance of 22 m. Calculate:
  a the time taken
  b the linear retardation of the vehicle
  c the angular velocity of the wheels at 50 km/h
  d the angular retardation of the wheels.

3 A 600 mm diameter pulley lifts a load at $0 \cdot 3$ m/s initially, and over a period of 10 s the speed is uniformly accelerated to $1 \cdot 4$ m/s. Determine:
  a the acceleration of the load
  b the angular acceleration of the pulley
  c the distance moved by the load
  d the number of turns made by the pulley.

4 A load is lifted by a cable which is wound round a drum of 180 mm diameter. The drum is initially running at 20 rev/min and is uniformly accelerated to 40 rev/min in 6 s. Determine:
  a the initial velocity of the load in m/s
  b the acceleration of the drum in rad/s^2
  c the acceleration of the load in m/s^2
  d the distance in m moved by the load during the 6 s.

5 **a** Give three formulae relating linear and angular velocity, linear and angular acceleration, and linear and angular distance, stating the units of each symbol used.
  b A flywheel 350 mm diameter is mounted on an axle which is 35 mm diameter. A string is wound round the axle and a weight is attached to the free end. The weight falls a distance of 1 m in 5 s. with uniform acceleration. Calculate:
    i the angular acceleration of the flywheel
    ii the final angular velocity of the flywheel
    iii the number of turns made by the flywheel and
    iv the final linear speed of a point on the rim of the flywheel.

6 A 150 mm diameter pulley has a rope wound round it. Starting from rest, 1 m of the rope is pulled off in 1 s. with uniform acceleration. Calculate:
  a the acceleration of the rope
  b the angular acceleration of the pulley
  c the final velocity of the pulley in rev/min
  d the number of revs made by the pulley while the rope is being pulled off.

7 A cylinder, starting from rest, rolls, without slipping, down an inclined plane with uniform acceleration. It covers a distance of $1 \cdot 3$ m in 2 ss. If the diameter of the cylinder is 100 mm, calculate:
  a its final linear velocity
  b its linear acceleration
  c its angular acceleration
  d the number of turns made by the cylinder.

8 a When an aircraft lands, its wheels are accelerated uniformly from rest (while skidding on the runway) to a maximum speed of 630 rev/min in $\frac{1}{2}$ second. Find:

  i the number of revolutions made by the wheels before the maximum speed of revolution is attained

  ii the angular acceleration of the wheels in rad/s.

  b Slipping between the wheels and the runway (as indicated by skid marks) ceases when the peripheral velocity of the wheels is equal to the linear velocity of the aircraft. Find:

  i the linear velocity of the aircraft in part **a** when the skid marks cease at 630 rev/min if the wheels are 1150 mm in diameter

  ii the length of the skid marks if the linear speed of the aircraft remains constant while the wheels are accelerating, i.e. skidding.

3.2   Newton's second law applied to angular motion

Consider the body shown in Fig. 3.1 rotating with constant angular acceleration about the pivot O. A small element of the body mass δm is a distance r from O. The *linear* acceleration of δm is then given by $a = r\alpha$, and its direction is perpendicular to the direction of r.

An acceleration is caused by an out-of-balance force, and so a force acts on δm, where

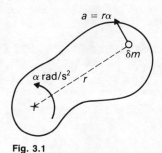

Fig. 3.1

Force on elemental mass $= \delta m \times a$ (using $P = ma$)
$$= \delta m \times r\alpha$$

The turning moment of this force about O $= \delta m \times r\alpha \times r$
$$= r^2\alpha\ \delta m$$

The total torque T is the sum of the moments of the forces on all the elements comprising the total mass

Therefore total torque, $T = \int r^2\alpha\ dm$

$$= \int r^2\ dm \times \alpha$$

but $\int r^2\ dm$ is the moment of inertia of the body about O (see Appendix 4).

Therefore $T = I\alpha$ Nm   (3.9)
where $I =$ moment of inertia in kg m^2
  $\alpha =$ angular acceleration in rad/s^2.

Equation (3.9) relates the torque applied to a body to the moment of inertia of the body and the angular acceleration produced. The equation is the angular equivalent of '$P = ma$' in linear motion.

Example 3.2

The moment of inertia of a flywheel is 100 kgm^2. Calculate the braking torque required to bring it to rest in 12 seconds from 150 rev/min.

$$\omega_1 = 150 \text{ rev/min} = \frac{150 \times 2\pi}{60} \text{ rad/s}$$

$$\text{therefore } \omega_1 = 15 \cdot 7 \text{ rad/s.}$$

$$\omega_2 = 0$$

Using $\omega_2 = \omega_1 + \alpha t$

$$\alpha = \frac{\omega_2 - \omega_1}{t} = -\frac{15 \cdot 7}{12} = -1 \cdot 3 \text{ rad/s}^2$$

i.e. a retardation of $1 \cdot 3$ rad/s^2

$$\begin{aligned}
\text{Braking torque} &= I\alpha \\
&= 100 \times 1 \cdot 3 \\
&= 130 \text{ Nm}
\end{aligned}$$

RADIUS OF GYRATION, k

The radius of gyration of a body is defined as the radius at which the mass of the body would have to be concentrated for the moment of inertia to remain unchanged. Thus, if the radius of gyration is denoted by k,

$$I = mk^2 \tag{3.10}$$

Example 3.3

The winding drum in a drive for a pit cage has a mass of 20 tonnes, is 10 m in diameter, and has a radius of gyration of 4 m. The total mass of the loaded cage and the rope is 6 tonnes and frictional effects may be taken as being equivalent to an additional force

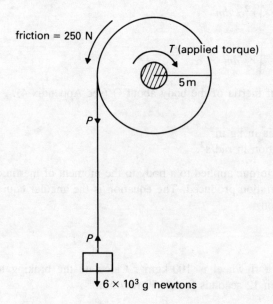

of 250 N acting tangentially to the circumference of the drum. Calculate the upward acceleration of the cage when a torque of $0 \cdot 4$ MNm is applied to the drum.

$$r = 5 \text{ m}, \ k = 4 \text{ m}$$

mass of drum = $20t = 20 \times 10^3$ kg
mass of cage = $6t = 6 \times 10^3$ kg

$$
\begin{aligned}
I \text{ for drum} &= mk^2 \\
&= 20 \times 10^3 \times 4^2 \\
&= 320 \times 10^3 \text{ kg m}^2
\end{aligned}
$$

Net torque used to turn drum $= T - 250 \times 5 - P \times 5$
where P is tension in rope.
Applying $T = I\alpha$ to drum

$$\text{or} \quad \alpha = \frac{T}{I} = \frac{0 \cdot 4 \times 10^6 - 1250 - P \times 5}{320 \times 10^3} \text{ rad/s}^2$$

$$= \frac{398 \ 750 - P \times 5}{320 \ 000}$$

but $a = r\alpha = 5\left(\dfrac{398 \ 750 - 5P}{320 \ 000}\right) = \dfrac{398 \ 750 - 5P}{64 \ 000}$

rearranging $P = \dfrac{398 \ 750 - 64 \ 000a}{5} = 79 \ 750 - 12 \ 800a$

Applying $P = ma$ to cage $\quad P - 6 \times 10^3 g = 6 \times 10^3 a$
$$79 \ 750 - 12 \ 800a - 6000g = 6000a$$
$$\text{whence } a = 1 \cdot 11 \text{ m/s}^2$$

Problems 3.2

1 **a** Prove that a rotating body of moment of inertia I, when acted upon by a constant torque T, undergoes an angular acceleration of magnitude T/I.

  b A marine engine drives a propeller steadily at 120 rev/min. The pitching of the ship causes the propeller to leave the water partially for $\frac{1}{2}$ s. If during this time the resistance torque of the propeller is reduced so that the unbalanced driving torque is 78 kNm, calculate the rev/min of the propeller at the end of the $\frac{1}{2}$ s period. The rotating parts have a mass of 10 Mg at a radius of gyration of $1 \cdot 2$ m.

2 A flywheel of $2 \cdot 5$ m diameter has a mass of 5 Mg and has a radius of gyration of 1 m. By means of a brake block applying a constant force on the rim, the flywheel is slowed down from 300 rev/min to 120 rev/min while the wheel turns through 140 rev. The coefficient of friction between the brake block and the flywheel rim is $0 \cdot 3$. Calculate:

  a the time taken to retard the flywheel
  b the torque required to produce the retardation
  c the force with which the brake block presses on the rim.

3 A bicycle front wheel may be regarded as a hoop of mass 2 kg and radius of gyration 0·3 m. It is set rotating freely about its axis with an angular velocity of 40 rad/s and is then brought to rest in 0·05 s by the application of the brakes.
   a If there are two brake blocks, each making contact with the wheel 0·3 m from the axis and the coefficient of friction is 0·25, what must be the contact force between each block and the rim?
   b What angle will the wheel turn through in coming to rest?

4 The loaded cage of a hoist has a mass of 2 tonne and is raised vertically by a rope which passes round a winding drum of 1· 2 m diameter; the other end of the rope is fastened to a balance weight and as the loaded cage is lifted the balance weight is lowered. The balance weight has a mass of 1·25 tonne; the drum has a mass of 0·6 tonne and has radius of gyration of 0·5 m. Find the torque that must be applied to the drum in order to accelerate the system when the loaded cage is being lifted with an acceleration of 1 m/s^2.

5 A cage which has a mass of 2 Mg hangs from a rope the upper end of which is wound round a drum 1·5 m in diameter. The drum has a mass of 1 Mg and has a radius of gyration of 0·6 m. If a steady torque of 18 kNm is applied to the drum shaft, calculate the acceleration of the cage, neglecting the weight of the rope.

6 A cage has a mass of 1·5 Mg and is raised by means of a rope coiled round a drum of 1·2 m diameter. The drum has a mass of 800 kg and has a radius of gyration of 0·5 m. The cage is accelerated from rest through a distance of 14 m in 7 s. Calculate:
   a the acceleration of the cage
   b the tension in the rope
   c the torque on the drum shaft.

7 Two masses of 10 kg and 8 kg are connected to the ends of a cord passing over a solid circular disc. The mass of the disc is 12 kg, its diameter is 0·6 m and its radius of gyration is 0·212 m. Determine, when the weights are moving freely:
   a the acceleration of the weights
   b the tensions in the two portions of the cord.
   The disc can be assumed to be mounted on frictionless bearings.

8 A shaft carries a flywheel of mass 4000 kg and radius of gyration 1 m. The drive to the shaft supplies 45 kW at 120 rev/min and the driving torque remains constant at all speeds. Calculate:
   a the bearing friction torque, assumed constant, if the system comes to rest from 120 rev/min with drive and brake off in 125 s
   b the braking torque necessary to bring the system to rest from 120 rev/min with the drive on in 20 s.

3.3　Motion in a circle with uniform angular velocity

CENTRIPETAL ACCELERATION

Consider a body moving in a circle of radius r with a constant angular velocity ω rad/s. The tangential velocity of the body at any instant will be v m/s where $v = r\omega$. After a small interval of time δt the body has moved through an angle $\delta\theta$. Although the magnitude of the velocity has remained unchanged, its direction certainly *has* changed,

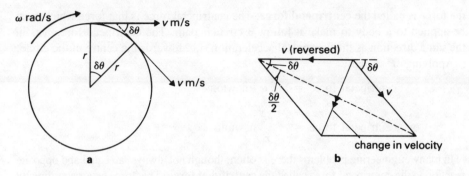

Fig. 3.2

the velocity having turned through an angle $\delta\theta$. Since velocity is a vector quantity, a change in direction is equivalent to a change in magnitude. Reference to the velocity diagram will show the magnitude and the direction of this change in velocity. Thus

$$\text{change in velocity} = 2v \sin \frac{\delta\theta}{2}$$

since $\delta\theta$ is small $\sin \dfrac{\delta\theta}{2} = \dfrac{\delta\theta}{2}$ in radians

therefore change in velocity $= 2v \dfrac{\delta\theta}{2}$

$$= v\delta\theta \text{ m/s}$$

A change in linear velocity in time δt is a linear acceleration. Thus

$$\text{linear acceleration} = \frac{\text{change in velocity}}{\text{time}}$$

$$= \frac{v\delta\theta}{\delta t}$$

In the limit, linear acceleration $= v \dfrac{d\theta}{dt}$

$$= v\omega$$

but $v = r\omega$, therefore linear acceleration $a = r\omega^2$ m/s^2 \hfill (3.11)

Alternatively, replacing ω by $\dfrac{v}{r}$ $\qquad a = \dfrac{v^2}{r}$ m/s^2 \hfill (3.12)

The linear acceleration derived in equations (3.11) and (3.12) is called the **centripetal acceleration**.

The centripetal acceleration *always* acts on a body moving in a circular path, and its direction is always *towards the centre of the circle*.

If a body experiences an acceleration then, using $P = ma$, there must be an out-of-balance force causing the acceleration. In the case of a body moving in a circular path, the force is called the **centripetal force**. The centripetal force is that force which must be applied to a body to make it follow a circular path. The centripetal force acts in the same direction as the centripetal acceleration, i.e. towards the centre of the circle.

Applying '$P = ma$',

$$\text{centripetal force} = m\,r\omega^2 \text{ newtons}$$

$$\text{or} \quad \text{centripetal force} = m\,\frac{v^2}{r} \text{ newtons}$$

In many engineering problems there is often, though not always, an equal and opposite reaction to the centripetal force called the **centrifugal force**. This force never acts directly on the body moving in a circle, but exists only as a reaction to the centripetal force. For example, consider a mass being swung in a circle at the end of string. The string is in tension, pulling both on the mass and on the pivot at the centre of the circle. The pull on the mass is the centripetal force, the pull on the pivot is the centrifugal force.

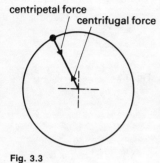

centripetal force
centrifugal force

Fig. 3.3

Example 3.4

The crank pin of an engine has a mass of 1 kg and its centre is 150 mm from the crank shaft centre. When the crankshaft is rotating uniformly at 2000 rev/min, calculate in magnitude and direction in each case:

  a the acceleration of the crank pin
  b the centripetal force on the crank pin
  c the centrifugal force on the crankshaft bearings

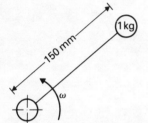

$$2000 \text{ rev/min} = \frac{2000 \times 2\pi}{60} \text{ rad/s}$$

therefore $\omega = 209\cdot4$ rad/s

a Since the crankpin moves in a circle of radius $0\cdot15$ m, the centripetal acceleration acts upon it is given by

$a = r\omega^2$

$\quad = 0\cdot15\,(209\cdot4)^2 = 6577\cdot3$ rad/s^2 towards centre of circle

b centripetal force $= mr\omega^2$

$\qquad\qquad\qquad\quad = 1 \times 6577\cdot3$

$\qquad\qquad\qquad\quad = 6577\cdot3$ newtons towards centre of circle

c centrifugal force $= 6577\cdot3$ newtons radially outwards from the centre of circle

Example 3.5

An object is hung at the end of a cord of length 1 m. Find the speeds in rev/min at which the object needs to rotate to keep the cord at an angle of 30° to the vertical.

Let T be the tension in the chord, and mg the weight of the object.
The object moves in a horizontal circle of radius r.

The centripetal force on the object is $T \sin 30°$

$$= T \times 0·5$$

but centripetal force $= mr\omega^2$

therefore $\quad T \times 0·5 = m \times 0·5 \times \omega^2$ **(1)**

Since there is no vertical displacement of the object, the forces vertically must be in equilibrium.

Therefore resolving vertically $T \cos 30° = mg$

$$\text{or} \quad T = \frac{2mg}{\sqrt{3}} \quad \textbf{(2)}$$

Substituting **(2)** in **(1)**

$$\frac{2mg}{\sqrt{3}} \times 0·5 = m \times 0·5 \times \omega^2$$

$$\text{or} \quad \omega^2 = \frac{2g}{\sqrt{3}} = 11·33$$

Therefore $\quad \omega = \sqrt{11·33} = 3·366$ rad/s

$$= \frac{3·366}{2\pi} \times 60 \text{ rev/min} = 32·1 \text{ rev/min}$$

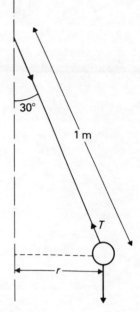

Example 3.6

The centre of gravity of a car is $0·6$ m above the ground and its track is $1·9$ m. Calculate the maximum speed at which the car can negotiate a horizontal bend of 60 m radius without overturning. What is the overturning moment at this speed if the car has a mass of 1000 kg?

The diagram shows the rear of the car entering the left-hand bend. If the car is *about* to overturn the forces acting on it are:

a its weight, mg, acting through the centre of gravity

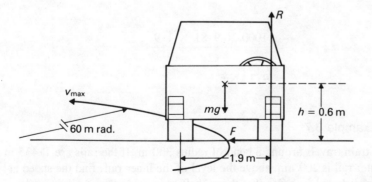

b the vertical reaction of the ground on the outside wheels only (since the inside ones are on the point of lifting)

c the frictional force (again acting on the outside wheels only), preventing the car skidding.

For equilibrium, resolving vertically $R = mg$.

Taking moments about the centre of gravity

$$F \times h = R \times \frac{t}{2}$$

where h is the height of centre of gravity above the ground, and $t = $ track width.

$$\text{Therefore } Fh = \frac{mgt}{2} \quad \left(Note\text{: The car will overturn if } Fh > \frac{mgt}{2} \right)$$

$$\text{Thus } F = \frac{mgt}{2h}$$

$$= \frac{m \times 9 \cdot 81 \times 1 \cdot 9}{2 \times 0 \cdot 6} = 15 \cdot 5 \, m \text{ newtons.}$$

F is the only force acting on the car radially, and so F is the centripetal force.

$$\text{Therefore } F = \frac{mv^2}{r}$$

$$\text{Substituting for } F \qquad 15 \cdot 5 \, m = \frac{mv^2}{r}$$

$$\text{therefore } \quad v^2 = 15 \cdot 5r = 15 \cdot 5 \times 60$$
$$\text{whence } \quad v = 30 \cdot 3 \text{ m/s}$$
$$= 109 \cdot 8 \text{ km/h}$$

Hence maximum velocity to avoid overturning $= 109 \cdot 8$ km/h

The overturning moment $= Fh$

$$= \frac{mgt}{2}$$

$$= \frac{1000 \times 9 \cdot 81 \times 1 \cdot 9}{2}$$

$$= 9 \cdot 3 \text{ kNm}$$

Example 3.7

A railway train travels around a bend of radius 300 m. If the rails are $1 \cdot 435$ m apart and the outer rail is 200 mm above the level of the inner rail, find the speed at which the train must travel in order that there shall be no side thrust on the rails.

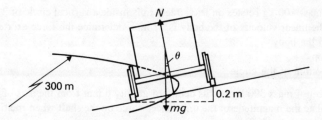

Let v be the speed the train must travel in order for there to be no side force on the rails. (If the bend were horizontal the side force on the rails would be the only way of obtaining a centripetal force.)

For a banked curve the centripetal force is supplied by the component of the normal reaction towards the centre of the curve.

If θ is angle of banking, then $\sin \theta = \dfrac{0 \cdot 2}{1 \cdot 435} = 0 \cdot 139$, giving $\theta = 8°$

For equilibrium of train

Resolving vertically $N = mg \cos 8°$

Applying '$P = ma$' horizontally $N \sin 8° = \dfrac{mv^2}{r}$

$$\text{therefore} \quad mg \cos 8° \sin 8° = \frac{mv^2}{300}$$

$$\text{therefore} \quad v^2 = 300g \cos 8° \sin 8°$$

$$\text{therefore} \quad v = 20 \cdot 14 \text{ m/s} = 72 \cdot 5 \text{ km/h}$$

Problems 3.3

1 a State what is meant by centripetal force and centrifugal force.

   b A 1 kg mass is attached to a cord $0 \cdot 6$ m long and it is whirled round in a vertical circle at a speed of 120 rev/min. Calculate the tension in the cord when the mass is:

     i at the top of the circle

     ii at the bottom of the circle.

   c Calculate the speed in rev/min at which the tension in the cord just disappears at the top of the swing.

2 A roundabout has a number of carriages hung from a rotating roof by means of chains $3 \cdot 5$ m long. The upper ends of the chains are attached to the circular roof of $3 \cdot 0$ m radius. Find the greatest speed in rev/min at which the roundabout can operate if the angle which the chains make with the vertical is not to exceed 20°.

3 A mass of 1 kg moving in a horizontal circle is kept in its path by means of a cord tied to a point 50 mm above the plane of rotation. The length of the cord is 500 mm. Determine:

   a the tension in the cord

   b the speed of rotation of the mass in rev/min.

4 A body of mass 100 kg rotates on frictionless rails inside a vertical circle of 3·3 m radius. If the linear velocity of the body is 27 m/s, determine the force exerted by the rails on the body
   a at the top
   b at the bottom of the circle.

5 A gear wheel of mass 300 kg has its centre of gravity 6 mm from the axis of rotation. Calculate the magnitude of the unbalanced force on the shaft when rotating at 300 rev/min.

6 a State what is meant by 'centripetal acceleration' and 'centripetal force'.
   b A uniform solid cylinder 600 m high and 50 mm diameter stands upright on the floor of a railway carriage which is travelling round a curve of radius 250 m. If the cylinder is just on the point of overturning calculate the speed of the carriage in km/h.

7 A light spring is threaded round a rod and secured to one end of it. The other end of the spring is attached to a 5 kg annular mass capable of sliding without friction along the rod. When stationary the centre of gravity of the weight is 0·3 m from the end of the rod to which the spring is attached. The rod is then rotated about this end in a horizontal plane at 40 rev/min. If the strength of the spring is 150 N/m find the radius of the circle of rotation of the 5 kg mass.

8 A uniform cylinder stands upright on a horizontal turntable. The turntable rotates about a vertical axis and the distance of the axis of the cylinder from the axis of the turntable is 0·6 m. The height of the cylinder is 150 mm and the diameter is 50 mm. The coefficient of friction between the cylinder and the turntable is 0·4. The speed of the turntable is increased until the cylinder either overturns or slides. State which occurs, and at what rev/min of the turntable.

9 A railway wagon of total mass 12 Mg, travels at a uniform speed of 70 km/h round a horizontal curve. The distance between the wheel tracks is 1·37 m and the centre of gravity of the wagon is on the vertical through the centre of the wheel base and 1·5 m above the ground. If the wagon is just on the point of overturning, determine the radius of the curve.

10 A centrifugal clutch consists of four blocks, each of mass 2 kg rotating in a housing and connected to the axis of rotation by springs of stiffness 10 N/mm. When stationary the centre of gravity of each weight is 125 mm from the axis of rotation and there is a 25 mm clearance between the blocks and the housing which is 325 mm internal diameter. Part of the arrangement is shown in the figure. Calculate:
   a the speed in rev/min at which the clutch will begin to transmit power
   b the force exerted by each block on the housing at 600 rev/min
   c the torque transmitted at 600 rev/min if the coefficient of friction between blocks and housing is 0·3.

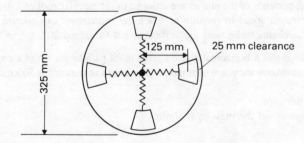

11 A lorry travels at 50 km/h round a horizontal curve of 200 m radius. The distance between the wheel tracks is $1 \cdot 45$ m; the centre of gravity of the vehicle is on the vertical through the centre of the wheel base and $1 \cdot 2$ m above the ground. The total mass of the vehicle is 8000 kg. Calculate:

    a the vertical force on the inner and outer wheels

    b the maximum speed in km/h at which the vehicle can negotiate the curve without overturning.

12 An aircraft of mass 4000 kg is flying in a horizontal circle of radius 400 m banked at an angle of $58° \ 30'$. Calculate:

    a the speed of the aircraft in knots

    b the total lift on the wings. (1 nautical mile $= 1852$ m)

13 A four-wheeled, two-axled vehicle travels at a uniform speed of $13 \cdot 4$ m/s round a horizontal curve of 100 m radius. The distance between the wheel tracks is $1 \cdot 5$ m, the centre of gravity of the vehicle is on the vertical through the centres of the wheel base and $1 \cdot 7$ m above the ground, and the total mass of the vehicle is 20 Mg. Determine:

    a the vertical force on each of the inner and outer wheels

    b the minimum radius of the horizontal curve that the vehicle can negotiate at $13 \cdot 4$ m/s without overturning (assume no skidding).

14 A motor car test track has a curve of radius 200 m banked at an angle of $15°$ to the horizontal.

    a Calculate the maximum speed possible of a car around the bend if no side thrust is to be felt on the tyre walls.

    b If the coefficient of friction between the tyres and the track is $0 \cdot 68$, calculate the maximum speed possible around the bend, assuming that no side slip takes place.

15 A car travelling at 180 km/h negotiates a banked curve of radius 500 m. Calculate the minimum angle of banking if the car is at the limit of adhesion with the road. The coefficient of friction between the tyres and the road is $0 \cdot 35$.

16 A railway engine of mass 10 tonnes travelling on tracks $1 \cdot 435$ m apart at 100 km/h negotiates a bend of radius 400 m. The outer track is 150 mm higher than the inner. Calculate the side thrust exerted by the rails on the outer wheels of the engine.

3.4 Simple harmonic motion

DEFINITION

A particle is said to move with simple harmonic motion (SHM) when its acceleration is directly proportional to its displacement from a fixed point in its path, and always directed towards that point.

A perfect example of SHM is the oscillation of a mass spring system. A spring of natural length l carries a mass of m kg. From its equilibrium position the mass is given a further downward displacement a and released. The mass will then oscillate vertically about the mean position as shown in the consecutive diagrams of Fig. 3.4. The mass moves with SHM since the restoring force in the spring, and hence the acceleration, is always directed towards the equilibrium position. The following relationship may be readily determined for the mass-spring system and shown to apply to SHM in general.

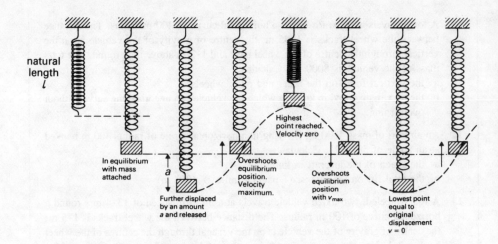

Fig. 3.4

a The restoring force is directly proportional to the displacement of the mass from the equilibrium position.

b The displacement of the mass can be timed to start from the moment that the mass is released (i.e. displacement $= a$ when $t = 0$), or from the moment the mass first passes through the equilibrium position (i.e. displacement $= 0$ when $t = 0$). We shall take the latter case.

c The velocity of the mass will be zero when the displacement is at its maximum ($\pm$), and at its maximum ($\pm$) when the displacement is zero. Thus when $t = 0$ at the equilibrium position the velocity wil be at its maximum.

d The acceleration of the mass will be directly proportional to the displacement (as is the restoring force). Thus at the equilibrium position the acceleration is zero, and at the maximum displacement ($\pm$) the acceleration is at its maximum ($\pm$). The acceleration will always be directed towards the equilibrium position.

Simple harmonic motion can also be represented by the movement of a point which is the projection on a diameter of a second point moving in a circle with constant angular velocity. For example, let P be a point moving in a circular path of radius a with constant angular velocity ω_n rad/s (Fig. 3.5). After time t seconds the line OP will have rotated through an angle of $\omega_n t$ radians. Q is the projection of P on to a horizontal diameter RS. As P rotates clockwise from ON, Q moves back and forth along RS with SHM. A graph of x, the displacement of Q from the origin O, against the angle that OP makes with ON produces a sinusoidal curve which repeats every 2π radians.

The displacement of Q at any time t is given by

$$x = a \sin \omega_n t \tag{3.13}$$

The velocity of Q at time t, (dx/dt), is equal to the horizontal component of v, the linear (tangential) velocity of P, i.e.

$$\text{velocity of Q} = \frac{dx}{dt} = v \cos \omega_n t$$

(*Note*: the velocity is in the same sense as the direction of measurement of x) but $v = \omega_n a$

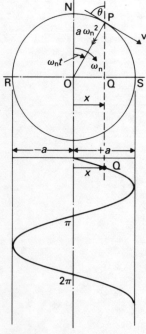

Fig. 3.5

therefore velocity of Q $= \omega_n a \cos \omega_n t$ (3.14)

but $a \cos \omega_n t = PQ = \sqrt{a^2 - x^2}$

therefore, numerically, the velocity of Q $= \omega_n \sqrt{a^2 - x^2}$ (3.15)

Equations (3.14) and (3.15) show that the velocity of Q has a maximum value of $\omega_n a$ when $x = 0$ i.e. when $\omega_n t = \pi/2$ or $3\pi/2$ radians. (*Note*: When $\omega_n t = \pi/2$ the velocity has a maximum positive value, and when $\omega_n t = 3\pi/2$ the velocity has a maximum negative value) i.e.

$$v_{max} = \pm \omega_n a \qquad (3.16)$$

P moves in a circular path and so will have a centripetal acceleration of $a\omega_n^2$ or v^2/a directed towards O. The horizontal component of this centripetal acceleration will be the acceleration of Q, d^2x/dt^2.

Thus acceleration of Q, $d^2x/dt^2 = -v^2/a \sin \omega_n t = -\omega_n^2 a \sin \omega_n t$ (3.17)

but $a \sin \omega_n t = x$, the displacement of Q from O.

Therefore acceleration of Q $= \dfrac{d^2x}{dt^2} = -\omega_n^2 x$ (3.18)

Equation (3.18) shows that the acceleration of Q is proportional to x and is at its maximum when $x = \pm a$ i.e. when $\omega_n t = 0$ or π radians. The negative sign indicates that the acceleration is opposite to the measured direction of x. Thus Q moves according to the definition of SHM.

Note: Equations (3.14) and (3.17) can be obtained directly from equation (3.13) by differentiation. Thus

$$x = a \sin \omega_n t$$

$$\frac{dx}{dt} = \omega_n a \cos \omega_n t$$

$$\frac{d^2x}{dt^2} = -\omega_n^2 a \sin \omega_n t = -\omega_n^2 x$$

The time for P to make one complete revolution, and also for Q to make one complete oscillation, is called the **periodic time** T.

Thus $T = \dfrac{\text{angular displacement in one revolution}}{\text{angular velocity}} = \dfrac{2\pi}{\omega_n}$ seconds   (3.19)

from equation (7.18) $\omega_n = \sqrt{\dfrac{\text{acceleration}}{\text{displacement}}}$

therefore $T = 2\pi \sqrt{\dfrac{\text{displacement}}{\text{acceleration}}}$ (3.20)

The radius a of the circle described by P is equal to the maximum displacement of Q from O and is called the **amplitude** of the vibration or oscillation.

The natural **frequency** f of the vibration is the number of oscillations made by Q in one second. If Q takes T seconds to perform one complete vibration in one second there will be $1/T$ vibrations, i.e.

$$f = \frac{1}{T}$$

$$= \frac{\omega_n}{2\pi} \text{ hertz (Hz)}$$

In this context ω_n is termed the **natural circular frequency** of the vibration.

Example 3.8

The periodic time of a body moving with SHM is 3 seconds and its amplitude is $0 \cdot 2$ m. Calculate:

    a its acceleration when 80 mm from the mean position
    b its velocity when 50 mm from the mean position

Periodic time, $T = \dfrac{2\pi}{\omega_n}$

therefore $\omega_n = \dfrac{2\pi}{T} = \dfrac{2\pi}{3}$ rad/s

a acceleration $\dfrac{d^2x}{dt^2} = -\omega_n^2 x$

$$= -\left(\frac{2\pi}{3}\right)^2 \times 0\cdot08$$

$$= 0\cdot35 \text{ m/s}^2$$

b velocity $= \omega_n\sqrt{a^2 - x^2}$

$$= \frac{2\pi}{3}\sqrt{(0\cdot2)^2 - (0\cdot05)^2}$$

$$= \frac{2\pi}{3}(0\cdot194) = 0\cdot405 \text{ m/s}$$

Example 3.9

A body moving with SHM has a velocity of $1 \cdot 3$ m/s at a distance of $1 \cdot 0$ m from the mean position, and a velocity of $1 \cdot 0$ m/s at a distance of $1 \cdot 3$ m from the mean position. Find:

a the amplitude of the motion

b the velocity at the mid-position

c the time elapsed between the instants when the body is moving at $1 \cdot 0$ m/s and when it is moving at $1 \cdot 3$ m/s.

a $v = \omega_n \sqrt{a^2 - x^2}$ thus $1 \cdot 3 = \omega_n \sqrt{a^2 - 1}$ **(1)**

and $1 \cdot 0 = \omega_n \sqrt{a^2 - 1 \cdot 3^2}$ **(2)**

dividing **1** and **2**

$$\frac{1 \cdot 3}{1 \cdot 0} = \frac{\cancel{\omega}_n \sqrt{a^2 - 1}}{\cancel{\omega}_n \sqrt{a^2 - 1 \cdot 69}}$$

squaring both sides

$$1 \cdot 69 = \frac{a^2 - 1}{a^2 - 1 \cdot 69}$$

$$(a^2 - 1 \cdot 69)1 \cdot 69 = a^2 - 1$$
$$1 \cdot 69 a^2 - 2 \cdot 856 = a^2 - 1$$
$$0 \cdot 69 a^2 = 1 \cdot 856$$
$$a^2 = 2 \cdot 69$$

Thus amplitude $a = 1 \cdot 64$ m

substituting in **2** $1 \cdot 0 = \omega_n \sqrt{1}$ therefore $\omega_n = 1$ rad/s.

b The velocity at the mid-position is the maximum velocity of the motion

$$v_{max} = a \omega_n$$
$$= 1 \cdot 64 \times 1 = 1 \cdot 64 \text{ m/s}$$

c The velocity at any instant is given by

$$v = a \omega_n \cos \omega_n t$$
therefore $1 \cdot 0 = 1 \cdot 64 \times 1 \times \cos t_1$ **(3)**
and $1 \cdot 3 = 1 \cdot 64 \times 1 \times \cos t_2$ **(4)**

from **3** $\cos t_1 = 0 \cdot 6098$ therefore $t_1 = 0 \cdot 9150$ s
from **4** $\cos t_2 = 0 \cdot 7927$ therefore $t_2 = 0 \cdot 6556$ s
Therefore time elapsed $= t_1 - t_2 = 0 \cdot 2594$ seconds

Example 3.10

The piston of an engine is assumed to move the SHM. The engine rotates at 300 rev/min, the crank is 200 mm long and the piston has a mass of 5 kg. Calculate:

    a the accelerating forces on the piston at the end of the stroke

    b the accelerating forces on the piston at quarter stroke

    c the velocity of the piston at quarter stroke.

a The natural circular frequency, $\omega_n = 300 \times \dfrac{2\pi}{60} = 10\pi$ rad/s

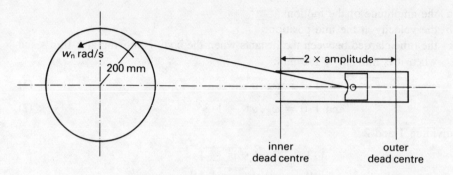

The distance between the inner and outer dead centres is twice the amplitude
$= 2 \times 0 \cdot 2 = 0 \cdot 4$ metres.

$$\text{Acceleration} = \frac{d^2x}{dt^2} = -\omega^2 x$$

The maximum acceleration occurs at the end of the stroke, i.e. when $x = a$.

$$\text{Therefore } \left(\frac{d^2x}{dt^2}\right)_{max} = -\omega_n^2 a$$
$$= -(10\pi)^2 0 \cdot 2 = -197 \cdot 39 \text{ m/s}^2$$

The accelerating force to give this acceleration is found by applying '$P = ma$'.

$$\text{i.e. } P = 5 \times 197 \cdot 39$$
$$= 986 \cdot 95 \text{ newtons}$$

b The stroke $= 2 \times$ amplitude

therefore $\frac{1}{4}$ stroke $= \frac{1}{2}$ amplitude   i.e. $\frac{1}{4}$ stroke $= 0 \cdot 1$ m.

$$\text{At } 0 \cdot 1 \text{ m } \frac{d^2x}{dt^2} = -\omega^2 x$$
$$= -(10\pi)^2 0 \cdot 1 = -98 \cdot 69 \text{ m/s}^2$$

Again applying '$P = ma$'
accelerating force $= 5 \times 98 \cdot 69 = 493 \cdot 45$ N

c Velocity at $\frac{1}{4}$ stroke (i.e. when $x = 0 \cdot 1$ m) is given by

$$v = \omega_n \sqrt{a^2 - x^2}$$
$$= 10\pi\sqrt{0 \cdot 2^2 - 0 \cdot 1^2}$$
$$= 10\pi\sqrt{0 \cdot 3} = 5 \cdot 44 \text{ m/s}$$

Example 3.11

A simple pendulum is constructed by attaching a mass of 5 kg to one end of a light string of length $0 \cdot 6$ m, the other end being rigidly fixed. The pendulum is displaced through a small angle θ from the vertical and released. Calculate the periodic time.

For a mass m kg and a length l, restoring force at maximum displacement

$$= mg \sin \theta$$
$$= mg \, \theta \text{ since } \theta \text{ is small}$$

Applying '$P = ma$' to mass

$$mg\theta = ma$$
$$\text{therefore } a = g\theta$$

Displacement of mass from mean position $= l \sin \theta$
$$\approx l\theta$$

From equation (3.20) $T = 2\pi \sqrt{\dfrac{\text{displacement}}{\text{acceleration}}}$

$$= 2\pi \sqrt{\frac{l\theta}{g\theta}}$$

$$= 2\pi \sqrt{\frac{l}{g}} \qquad\qquad \textbf{(1)}$$

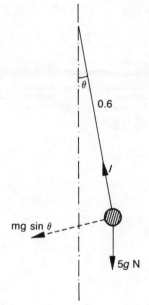

Since $l = 0 \cdot 6$ m, then $T = 2\pi \sqrt{\dfrac{0 \cdot 6}{9 \cdot 81}} = 1 \cdot 554$ seconds.

Notice that in equation **1** the periodic time of a simple pendulum is dependent only on the length of the string, and is independent of the mass attached to the end provided that the angle of displacement is such that

$$\sin \theta = \theta$$

FORCED VIBRATIONS

So far we have seen that a system which is set vibrating by an initial disturbance will vibrate with its own natural frequency. Once the initial disturbance is over there is no further disturbance and the system is said to vibrate freely. These free vibrations will eventually die out because of internal damping effects which cause the amplitude of each successive oscillation to be reduced.

However there are many engineering situations in which a system is forced to vibrate through contact with an adjacent system which is vibrating. The forced vibrations induced in the original system do not die out, but continue for as long as the disturbing force is present.

There are many familiar examples of forced vibrations: the vibrations caused in the seats, panels and windows of a bus due to the vibration of the engine, particularly at idling speeds; gear level 'rattle' in a motor car; the vibrations of bridge structure, particularly of the suspension type, due to the rumble of traffic over it. In fact wherever there is rotating or reciprocating machinery there may be vibrations forced upon neighbouring structures and plant. Usually this will not be much of a problem beyond being a nuisance, and steps can be taken to reduce the effects to a minimum. However, in certain circumstances and conditions the forced vibration can produce undesirable and damaging effects. It is these conditions that will now be examined in more detail.

RESONANCE

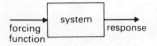

Fig. 3.6

As already stated all systems have a natural frequency of vibration. When a system is subjected to a harmonic disturbing force, the kind that might be produced by rotating machinery, then the system too will be caused to vibrate in a harmonic fashion. The harmonic disturbing force is called the **forcing function** and the manner in which the system reacts is called the **response**. The forcing function can be regarded as the input to the system and the response as the output from the system (Fig. 3.6). The important relationships between the forcing function and the response are as follows:

a The frequency of the response is equal to the frequency of the forcing function.
b The amplitude of the response will depend on the magnitude of the forcing function and the similarity between the **forcing frequency** and the **natural frequency** of the system. If the forcing frequency differs from the natural frequency the system will respond only moderately to the forcing function. If, however, the forcing frequency is close to the natural frequency then the system will respond readily to the forcing function and vibrations with large amplitudes will result. If the forcing frequency is equal to the natural frequency then a state of **resonance** occurs where, in theory, infinite amplitudes are possible, although in practice there is always some damping which limits the amplitude. The forcing frequency at resonance is called the **resonant frequency**. The resonant frequency is then equal to the natural frequency of the system.
c Resonance occurs when the phase difference between forcing function and the response is $\pi/2$ radians, that is the response **lags** the forcing function by $\pi/2$ rads.

When the forcing frequency is much less than the natural frequency of the system the two vibrations are almost exactly in phase, but the phase difference increases as resonance is approached as shown in Fig. 3.7. As the forcing frequency increases beyond the natural frequency the phase difference increases until at large differences in frequency the phase difference is π radians or half a cycle.

Fig. 3.7

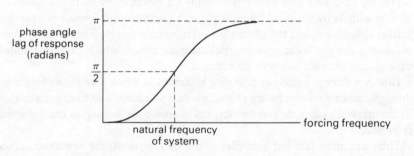

Resonance can be demonstrated by two simple pendulums A and B of equal length, so that their natural frequencies are the same, attached to the same flexible support such as a taut wire or string (Fig. 3.8). Initially pendulum B is at rest, and pendulum A is set vibrating with small amplitude. A then provides a forcing function of frequency equal to the natural frequency of B and as a result B will begin vibrations with the same frequency, and lagging $\pi/2$ or $\frac{1}{4}$ cycle behind A. The amplitude of vibrations of B will increase until it becomes the dominant influence and the amplitude of A will

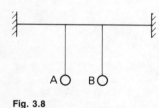

Fig. 3.8

diminish to eventually reach zero. The process is then reversed with B acting the role of the forcing function and A the response. The motion continues until natural damping (such as provided by air resistance) eventually stops the system vibrating altogether.

A resonant condition causes tool chatter in machining operations, the squeal of brakes on a car, etc. It can be extremely dangerous, and failure of components and structures can take place.

Problems 3.4

1   A body of mass 50 kg is oscillating in a horizontal straight line with simple harmonic motion. The amplitude of the oscillation is $1 \cdot 3$ m and the periodic time of a complete oscillation is 10 s. Calculate:
   a   the maximum force acting on the body
   b   the velocity of the body when it is $0 \cdot 3$ m from the centre of oscillation
   c   the time taken to travel $0 \cdot 3$ m from an extreme point of the oscillation.

2   A body of mass 10 kg moves with SHM with a frequency of 75 oscillations per minute. At maximum displacement the acceleration of the body is 16 m/s^2. Calculate:
   a   the amplitude of the oscillation
   b   the displacement of the body from the mid-position when the force acting on the body is 75 N.

3   A valve of mass 8 kg moves horizontally with SHM over a stroke of $0 \cdot 15$ m at the rate of 4 double strokes (i.e. backwards and forwards) per second. Events controlled by the valve occur when it is 40 mm from its mid-position and moving towards it, and again when it is 25 mm beyond its mid-position on the same stroke. Determine:
   a   the time between the events
   b   the maximum force to drive the valve (ignoring frictional resistance).

4   A body of mass 25 kg moves with simple harmonic motion with a frequency of 100 complete oscillations per minute. The maximum velocity attained by the body is 5 m/s. Determine:
   a   the amplitude of the motion
   b   the displacement of the body from the mid-position when the force acting on the body is 750 N
   c   the time taken by the body to move from the extreme position to a point midway between the extreme and the mid-position of the motion.

5   A 5 kg mass oscillates with simple harmonic motion of amplitude $0 \cdot 25$ m and a frequency of 60 oscillations per minute. Determine:
   a   the maximum force acting on the body during the oscillation
   b   the distance moved by the body during the first $0 \cdot 125$ s from the extremity of the oscillation
   c   the velocity at the position given in b.

6   A mass of 25 kg is attached to the end of an elastic spring of natural length $0 \cdot 25$ m whose other end is fixed. The system is in equilibrium when the spring is $0 \cdot 3$ m long. The weight is then pulled down a further 25 mm and released. Find:
   a   the periodic time of the resulting motion
   b   the maximum acceleration of the weight.

7 A helical spring with a free length of 250 mm has one end fixed and a weight of 6 kg attached to the other end. The spring and the weight hang vertically. The weight is pulled down until the length of the spring is 320 mm and then released. If the weight makes 195 complete oscillations per minute, determine:
  a the strength of the spring in N/m
  b the amplitude of the motion in mm.

8 **a** A 1·5 kg mass is hung on the end of a helical spring and is set vibrating vertically. The mass makes 108 complete oscillations per minute. Determine the stiffness of the spring in newtons per metre.
  b If the maximum extension of the spring during the vibrations is 125 mm calculate the amplitude of the vibration.

9 When a certain load is suspended from one end of a spring of stiffness 400 N/m and caused to perform vertical SHM, the number of complete oscillations per minute is 76. Calculate:
  a the value of the load
  b the acceleration of the weight and the tension in the spring when it is 25 mm below the mean position.

10 A 2 kg mass is hung on the end of a helical spring and is set vibrating vertically. The weight makes 100 complete oscillations in 55 s. Determine:
  a the strength of the spring
  b the maximum amplitude of vibration in mm if the weight is not to leave the hook during the motion.

11 A spring of stiffness 60 N/m is suspended vertically and two equal weights of 80 N each are attached to the lower end. One of these weights is suddenly removed and the system oscillates. Determine:
  a the periodic time of the oscillation
  b its amplitude
  c the velocity and acceleration of the weight when passing through the half amplitude position.

12 A pendulum clock keeps correct time at a place where $g = 9·808$ m/s^2. How many seconds will it gain or lose in 24 hours if taken to a place where $g = 9·812$ m/s^2?

3.5   Kinetic energy of rotating bodies

In Section 1·5 we derived an expression for the linear kinetic energy of a body of mass m kg moving with a velocity v m/s thus:

$$\text{linear kinetic energy} = \tfrac{1}{2}mv^2 \text{ joules}$$

This linear kinetic energy is sometimes referred to as the kinetic energy of translation, implying that the centre of gravity of the body is changing its position. A body that is purely rotating about any given point has motion but no kinetic energy of translation. The energy possessed by a body moving with a pure rotation about a fixed point is called the angular kinetic energy, or kinetic energy of rotation of the body. For example the body in Fig. 3.9 is rotating about the fixed point O with angular velocity ω rad/s. Consider small element of the body of mass δm at a distance r from O. If v m/s is the linear velocity of the elemental mass then,

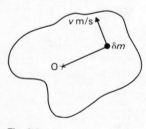

Fig. 3.9

kinetic energy of the element $= \frac{1}{2} \delta m v^2$

$$= \frac{1}{2} \delta m (r\omega)^2$$

The total kinetic energy of the body is equal to the sum of the kinetic energies of all such elements that comprise the body, i.e.

$$\text{total KE} = \int \frac{1}{2} r^2 \omega^2 \, dm$$

$$= \frac{1}{2} \omega^2 \int r^2 \, dm$$

Now $\int r^2 \, dm$ is the **moment of inertia**, I, of the body about O (see Appendix 4).

Therefore angular kinetic energy $= \frac{1}{2} I \omega^2$ joules   (3.21)

(The dynamic equivalence of *angular* KE $= \frac{1}{2} I \omega^2$ and *linear* KE $= \frac{1}{2} m v^2$ should be readily recognised.)

WORK DONE BY A TORQUE

Any change in the angular kinetic energy of a body must be as a result of an applied torque. The work done by a constant torque is equal to the product of the torque T, and the angular displacement θ.

work done by torque = change in angular kinetic energy

$$\text{or} \quad T\theta = \frac{1}{2} I (\omega_2^2 - \omega_1^2)$$ (3.22)

If the torque is variable during the angular kinetic energy change, equation (3.22) is modified to read

$$\int_{\theta_1}^{\theta_2} T \, d\theta = \frac{1}{2} I (\omega_2^2 - \omega_1^2)$$ (3.23)

The integral on the left can be determined provided that T as a function of θ is known, or by computing the area under the graph of T against θ between θ_1 and θ_2.

A body may possess both linear and angular kinetic energies. Thus

total KE of system = linear KE + angular KE

The laws of conservation of energy as used previously will still apply.

Example 3.12

A truck has a total mass of 4 tonnes. Each of the two axles with its wheels has a mass of 500 kg and a radius of gyration of $0 \cdot 3$ m. The diameter of each of the wheels is $0 \cdot 85$ m. If the truck is initially moving at 10 m/s, calculate:
  a the total KE of the truck
  b the distance travelled along a level track against a resistance of 200 N.

a Total KE = linear KE + angular KE

$$\text{linear KE} = \tfrac{1}{2}mv^2$$

$$= \tfrac{1}{2}(4 \times 10^3) \times 10^2 = 200 \text{ kJ}$$

angular KE of 2 wheels and axles $= 2(\tfrac{1}{2}I\omega^2)$

$$= 2\left[\tfrac{1}{2}(mk^2)\left(\frac{v}{r}\right)^2\right]$$

$$= 2\left[\tfrac{1}{2} \times 500 \times (0\cdot3)^2 \times \frac{10^2}{0\cdot425^2}\right] = 24\cdot9 \text{ kJ}$$

Total KE $= 200 + 24\cdot9$
$= 224\cdot9 \text{ kJ}$

b Work done against resistance $=$ change in KE
If x is the distance travelled
then $200\,x = 224\cdot9 \times 10^3$
therefore $x = 1124\cdot5 \text{ m}$

FLYWHEELS

If a machine operated by an engine or motor requires a steady unvarying supply of energy a flywheel is fitted which will store energy (angular KE) when the motor's output is excessive and give up this stored energy when the output falls. A flywheel is thus a reservoir of energy.

Example 3.13

A rolling mill requires an average of 300 kW during the 10 seconds that the billet is passing through the mill. An electric motor whose constant output is 255 kW drives the mill, and during the operation the flywheel speed falls from 80 to 72 rev/min. Find:
a the moment of inertia of the flywheel
b its mass, if its radius of gyration is 3 m

$$80 \text{ rev/min} = \frac{80 \times 2\pi}{60} \text{ rad/s} = \frac{8\pi}{3} = 8\cdot378 \text{ rad/s}$$

$$72 \text{ rev/min} = \frac{72 \times 2\pi}{60} \text{ rad/s} = \frac{12\pi}{5} = 7\cdot540 \text{ rad/s}$$

If I is the moment of inertia of the flywheel, loss in angular kinetic energy

$$= \tfrac{1}{2}I(\omega_1^2 - \omega_2^2)$$

$$= \tfrac{1}{2}I(8\cdot378^2 - 7\cdot540^2)$$

This loss is equal to the extra energy required

$$= (300 - 255) \times 10^3 \text{ joules per second}$$

$$= 45 \times 10^3 \text{ J/s}$$

therefore $\frac{1}{2}I(8 \cdot 378^2 - 7 \cdot 54^2) = 45 \times 10^3$

$$I = 90 \times 10^3 / 8 \cdot 378^2 - 7 \cdot 54^2 = 6747 \text{ kg m}^2$$

Now $I = mk^2$

$$\text{therefore } m = \frac{I}{k^2} = \frac{6747}{3^2} = 749 \cdot 7 \text{ kg}$$

Example 3.14

A cord is wound around a winding drum of $0 \cdot 5$ m diameter and moment of inertia 25 kgm^2. A mass of 50 kg is attached to the free end of the cord and is lifted from rest to a height of 20 m by applying a constant torque of 200 Nm to the drum. Calculate:

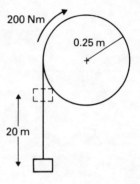

  a the work done by the torque in lifting the mass
  b the time taken
  c the kinetic energy of the drum after 20 m
  d the power developed after 20 m.

a Work done by torque = $T\theta$ joules
Angle turned through by drum in raising mass 20 m = $20/0 \cdot 25 = 80$ radians
Therefore work done by torque = 200×80
$$= 16 \text{ kJ}$$

b Work done by torque = KE of drum + KE of mass + potential energy of mass.

$$\text{Therefore } T\theta = \tfrac{1}{2}I\omega^2 + \tfrac{1}{2}mv^2 + mgh$$

where ω and v are the angular and linear velocities after 20 m respectively.

$$\text{Thus } 16 \times 1000 = \tfrac{1}{2} \times 25 \times \frac{v^2}{0 \cdot 25^2} + \tfrac{1}{2} \times 50 \times v^2 + 50 \times 9 \cdot 81 \times 20$$

$$16\,000 = 200v^2 + 25v^2 + 9810$$
$$v^2 = 27 \cdot 5$$
$$v = 5 \cdot 245 \text{ m/s}$$

$$\text{using } s = \frac{u + v}{2} \times t$$

$$20 = \frac{5 \cdot 245}{2} \times t \quad \text{therefore } t = 7 \cdot 63 \text{ seconds.}$$

c Kinetic energy of drum $= \tfrac{1}{2}I\omega^2$

$$= \tfrac{1}{2} \times I \times \left(\frac{v}{r}\right)^2$$

$$= \tfrac{1}{2} \times 25 \times \frac{27 \cdot 5}{0 \cdot 0625}$$

$$= 5500 \text{ J}$$
$$= 5 \cdot 5 \text{ kJ}$$

d Power = rate of doing work

$$= \frac{d}{dt}(T\theta) \qquad = T\frac{d\theta}{dt} \text{ since } T \text{ is constant}$$

$$= T\omega$$

$$= T\frac{v}{r}$$

$$= \frac{200 \times 5\cdot245}{0\cdot25} = 4196 \text{ watts}$$

$$= 4\cdot196 \text{ kW}$$

Problems 3.5

1 The flywheel of a power press has a mass of 60 kg and has a radius of gyration of 0·25 m. The flywheel is rotating at 500 rev/min when the ram is operated. The average force on the ram is 50 kN and it punches out a hole in material 10 mm thick.
   a Calculate the reduction in speed of the flywheel during the punching-out process.
   b Find the number of revolutions made by the flywheel if this speed reduction takes place in 2 seconds.

2 A punching machine has a flywheel of mass 200 kg and a radius of gyration of 0·3 m and it is running at 300 rev/min. During a certain punching operation through 5 mm plate a force of 200 kN is required on the punch.
   a Calculate the new speed of the flywheel just after the operation is completed, assuming that no energy is supplied from the motor during the operation.
   b If the flywheel must now be run up to its original speed in 5 s, what torque must be applied to it from the motor?

3 A machine press is worked by an electric motor delivering 2·5 kW continuously. At the commencement of the operation a flywheel of mass 370 kg and radius of gyration 0·3 m on the machine is rotating at 240 rev/min. The pressing operation requires 5 kJ of energy and occupies 1·0 s. Determine the reduction in the rev/min of the flywheel after each pressing.

4 **a** A shearing machine is fitted with a flywheel of mass 2000 kg and with a radius of gyration of 1·0 m. If the wheel reaches a speed of 210 rev/min in 2 minutes from rest calculate the steady torque required to attain this speed.
   b If the work done in shearing a piece of steel amounts to 20 000 joules and all the energy required for the operation comes from the flywheel, calculate the rev/min to which the flywheel will be reduced.

5 A torpedo is driven by expending the energy stored in a flywheel. The weight of the flywheel is 100 kg and the radius of gyration is 0·25 m. If the initial speed of the flywheel is 10 000 rev/min, find its speed in rev/min after the torpedo has run 500 metres assuming the average resistance to motion is 3·5 kN.

6 The total mass of a truck, including its four wheels, is 14 tonnes. Each wheel has a mass of 150 kg and has a diameter of 1·2 m and a radius of gyration of 0·32 m. Calculate how much work is done on the truck on a level track in increasing its velocity uniformly from 25 km/h to 35 km/h. Determine also the average forces causing the acceleration if the truck moves 400 metres during the operation. (Frictional resistances can be ignored.)

7 A truck body of mass 1 tonne is carried by four solid disc wheels which roll without slipping along a horizontal track. Each wheel weighs 150 kg and is 1·0 m diameter. The truck has a velocity of 1·3 m/s when it strikes a spring which is initially compressed 0·15 m. Determine the strength of the spring in N/m if the truck is brought to rest by compressing the spring by an additional 0·08 m.

$$\left(k = \frac{R}{\sqrt{2}}\right)$$

8 A container, A, of mass 300 kg runs on an overhead rail. The container is mounted on two wheels each weighing 65 kg and having a diameter of 1·1 m and a radius of gyration of 0·4 m. The assembly is moving to the left at a velocity of 1·0 m s, when it strikes a spring B of strength 600 N/m. Determine the amount in cm that the spring is compressed in bringing the assembly to rest.

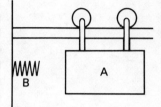

4 Stress and strain

The contents of this chapter will enable you to apply the basic principles of stress and strain analysis to complex stress situations in two dimensions. So far in this book we have made the assumption that the physical properties of a material, in particular its elastic constants E (Young's modulus) and ν (Poisson's ratio), are the same in all directions at a point in the material. Such a material is defined as being **isotropic**. However not all materials are isotropic. In some cases (e.g. aggregates) the value of the elastic constants depends on the direction in which they are measured, and in these instances the material is said to be **anisotropic**. In anisotropic materials there are up to 21 independent elastic constants. When the elastic constants are different in three mutually perpendicular planes the material is said to be **orthotropic**. Timber is an example of an orthotropic material.

The mathematical treatment of orthotropic and anisotropic materials is beyond the scope of this book, and we will confine ourselves to the analysis of isotropic materials only.

4.1 Stresses in thin-walled cyclinders

A thin-walled cylinder is defined as one in which the ratio of wall thickness to diameter does not exceed approximately 1 : 20. In this case the stresses due to an internal fluid pressure are assumed to be uniformly distributed through the cylinder wall. Figure 4.1 shows such a thin-walled cylinder subjected to an internal pressure p, which acts equally in all directions.

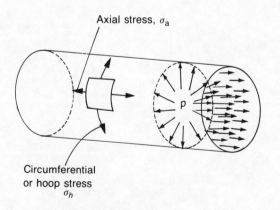

Axial stress, σ_a

Circumferential or hoop stress σ_h

Fig. 4.1

The stresses that result are; a longitudinal or axial stress σ_a, which results from a tendency to stretch the cylinder longitudinally due to the pressure acting on the circular ends (Fig. 4.2a) and a circumferential or hoop stress, σ_h which results from a tendency to increase the diameter of the cylinder due to the pressure acting radially on the inside curved surface (Fig. 4.2b).

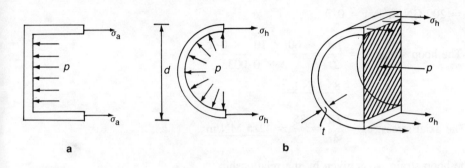

a   b   Fig. 4.2

Considering the equilibrium of the *forces* exerted in each case

(i)   Tensile force = Internal pressure × area of end plate
                    = axial stress × area of annular cross-section

i.e. $p \times \dfrac{\pi d^2}{4} \simeq \sigma_a \times \pi dt$

where p = internal pressure in N/m^2
      d = internal diameter of cylinder
      t = wall thickness
   πdt = approximate area of annular cross-section
(i.e. approx circumference × thickness) given that wall thickness is small in relation to diameter. Re-arranging

$$\sigma_a = \frac{pd}{4\pi t} \qquad\qquad\qquad (4.1)$$

(ii)   Internal pressure × projected area
                    = hoop stress × area of wall cross-section of length l
       $p \times d \times l = \sigma_h \times 2t \times l$

Re-arranging

$$\sigma_h = \frac{pd}{2t} \qquad\qquad\qquad (4.2)$$

Note that there is no approximation in assuming that the pressure acting radially around the inside of a semi-circular segment is equivalent to the pressure acting on the projected area ($d \times l$) because the curved area resolves to the shaded area. Equations (4.1) and (4.2) show that the hoop stress is always twice the longitudinal stress, and thus failure of the cylinder due to internal pressure will always be in an axial direction.

Example 4.1

A thin, mild steel cylinder is $0 \cdot 3$ m long, 100 mm diameter, and has a wall thickness of $3 \cdot 5$ mm. Calculate the value of the hoop and axial stresses when the cylinder is subjected to an internal pressure of 60 bar. Calculate also the change in diameter of the cylinder in the stressed condition.

$E = 200$ GN/m^2, $\nu = 0 \cdot 3$

$$\text{The hoop stress, } \sigma = \frac{pd}{2t} = \frac{60 \times 10^5 \times 0 \cdot 1}{2 \times 0 \cdot 0035}$$

$$= \underline{85 \text{ MN/m}^2}$$

$$\text{The axial stress, } \sigma_a = \frac{pd}{4t} = \frac{85}{2} = \underline{42 \cdot 5 \text{ MN/m}^2}$$

The hoop strain, ϵ_h is given by the relationship

$$\epsilon_h = \frac{\sigma_h}{E} - \nu \frac{\sigma_a}{E}$$

and since the circumference $= \pi \times$ diameter
then hoop strain, $\epsilon_h =$ diametral strain, ϵ_d

$$\text{hence } \epsilon_d = \frac{85 \times 10^6}{200 \times 10^9} - \frac{0 \cdot 3 \times 42 \cdot 5 \times 10^6}{200 \times 10^9}$$

$$= 3 \cdot 6 \times 10^{-4}$$

$$\therefore \text{ The change of diameter } = 3 \cdot 6 \times 10^{-4} \times 100$$

$$= \underline{0 \cdot 036 \text{ mm}}$$

Example 4.2

Part of an air compressor consists of a cylinder with closed ends and diameter $0 \cdot 75$ m. The yield stress of the material of the cylinder is 760 MN/m^2 and a safety factor of 4 is to be used. If the cylinder is internally pressurised to 50 bar calculate the required wall thickness.

For design purposes only the hoop stress needs to be considered, as it is that which reaches the limiting value first.

$$\text{Hence maximum } \sigma_h = \frac{760 \times 10^6}{4} = \underline{190 \times 10^6 \text{ N/m}^2}$$

$$\text{Since } \sigma_h = \frac{pd}{2t} = 190 \times 10^6 \text{ N/m}^2$$

$$\text{then, } t = \frac{50 \times 10^5 \times 0 \cdot 75}{2 \times 190 \times 10^6} = 0 \cdot 0098$$

The required wall thickness is $9 \cdot 8$ mm.

Example 4.3

A laminated pressure vessel consists of two thin co-axial cylinders. The outer cylinder is a 'shrink-fit' on the inner one, and the mean diameter of the assembly is 250 mm. The thickness of the outer cylinder is $2 \cdot 5$ mm and that of the inner 2 mm. Prior to 'shrink-fitting' the inner diameter of the outer cylinder was $0 \cdot 2$ mm larger than the outer diameter of the inner cylinder, i.e. the interference was $0 \cdot 2$ mm. Calculate the stresses in each cylinder due to the assembly. $E = 200$ GN/m^2. Clearly a pressure p acts at the interface of the two cylinders which tends to increase the diameter of the outer cylinder and decrease the diameter of the inner one.

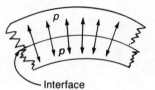

The sum of the increase and decrease is equal to the initial difference in diameter, i.e. $0 \cdot 2$ mm. Using the relationship deduced in Example 4.1, and since no longitudinal stresses exist in the problem:

$$\epsilon_d = \sigma_h/E$$

Thus, the change in diameter of the outer cylinder

$$= \sigma_h/E \times \text{diameter}$$

$$= \frac{pd^2}{2tE}$$

$$= \frac{p \times 0 \cdot 25^2}{2 \times 0 \cdot 0025 \times 200 \times 10^9} = 6 \cdot 25 \times 10^{-11} p \text{ metres}$$

and the change in diameter of the inner cylinder

$$= \frac{pd^2}{2tE}$$

$$= \frac{p \times 0 \cdot 25^2}{2 \times 0 \cdot 002 \times 200 \times 10^9} = 7 \cdot 8 \times 10^{-11} p \text{ metres}$$

Adding these changes gives

$$(6 \cdot 25 \times 10^{-11} + 7 \cdot 8 \times 10^{-11}) p = 0 \cdot 0002$$
$$\text{whence } p = 1 \cdot 42 \text{ MN/m}^2$$

Problems 4.1

1   Calculate the circumferential and longitudinal stresses induced in a thin-walled cylinder 100 mm diameter and wall thickness $4 \cdot 5$ mm when subjected to an internal pressure of 20 bar.

2 Calculate the safe internal working pressure for a thin-walled cylindrical pressure vessel if the maximum allowable stress is 160 MN/m^2. The cylinder is 0·6 m diameter and has a wall thickness of 3 mm.

3 A submarine hull may be regarded as a long cylindrical tube of mean diameter 8·5 m. If the safe working stress is limited to 100 N/mm^2 and the maximum design gauge pressure on the outside of the hull due to immersion at depth is 2 MN/m^2, calculate the thickness of the hull wall required.

4 A thin-walled pressure vessel in the form of a cylinder with ends closed by flat plates is subjected to an internal pressure of 6·5 MN/m^2. The vessel has a diameter of 0·35 m and a wall thickness of 2·75 mm. Calculate the change in diameter and length, and hence volume, of the cylinder that occurs between the stressed and unstressed conditions. $E = 200$ GN/m^2, $\nu = 0·3$

5 A pressurised cylindrical storage tank is constructed from welded steel sheet 8 mm thick. The diameter of the tank is 2·2 m and it is 4·5 m long. The elastic limit stress of the steel is 280 N/mm^2 and a safety factor of 3 is to be incorporated into the calculations. The welded seams of the tank are regarded as being only 80 per cent as strong as the solid sheet. Calculate the maximum internal pressure that can be used in the tank, and the change in volume that takes place when the tank is pressurised.

Take $E = 208$ GN/m^2 and $\nu = 0·3$.

6 A laminated cylinder consists of a thin steel cylinder 'shrunk' onto an aluminium one. The steel cylinder is 4 mm thick and the aluminium cylinder is 2·5 mm thick. The mean diameter of the assembly is 120 mm, and the interference prior to assembly is 0·01 mm. Calculate the stresses in each shell caused by the assembly. E for steel is 208 GN/m^2, and E for aluminium is 72 GN/m^2.

7 A hydraulic cylinder consists of a steel sleeve of 3·5 mm wall thickness which is shrunk onto a copper tube of wall thickness 2 mm. The mean diameter of the cylinder is 75 mm and when pressurised with hydraulic fluid it operates a ram which exerts a force of 250 kN. Calculate the stresses in the steel and copper due to the combination of shrink fitting and internal pressure. Prior to assembly the interference fit of the cylinder was 0·1 mm, E for steel is 208 GN/m^2, and E for copper is 80 GN/m^2.

4.2 Complex stress and strain

In a simple one-dimensional system it has been shown that the application of a direct stress will produce a direct strain, and that the application of a shear stress will produce a shear strain.

Even in these simple cases we were aware that stresses other than the applied ones acted in other directions to those specified. For example consider an example of simple tension in which an applied tensile stress σ_x, is applied to a bar of material (Fig. 4.3).

If we consider the stresses that act on a plane of the bar inclined at an angle θ to the axial direction it can be seen that for the portion of the beam (say) to the right of this plane to be in equilibrium a direct stress σ_x must act on the plane (Fig. 4.4a).

The direct stress σ_x that acts on the inclined plane can be resolved into two components; one parallel and one perpendicular to the plane. The perpendicular component is a direct stress σ, and the parallel component is shear stress τ.

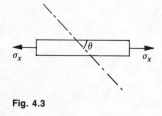

Fig. 4.3

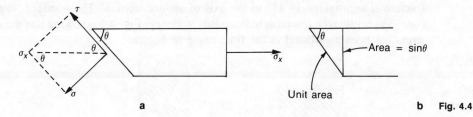

a   **b**   Fig. 4.4

If we assume, for the sake of convenience, that the inclined plane is of unit cross-sectional area, then the true cross-sectional area of the bar is $\sin \theta$, (Fig. 4.4b). The tensile force acting on the plane is thus $\sigma_x \sin \theta$ and the resolved part of this force perpendicular to the plane is $\sigma_x \sin \theta \times \sin \theta$.

For the equilibrium of *forces*, resolving perpendicularly to the plane gives:

$$\sigma \times 1 = \sigma_x \times \sin \theta \times \sin \theta$$

$$\begin{array}{ccc}\text{(stress} \times \text{area} & = \text{(stress} \times \text{area} & \text{(resolved perpendicular} \\ = \text{force)} & = \text{force)} & \text{to plane)}\end{array}$$

$$\sigma = \sigma_x \sin^2 \theta \tag{4.3}$$

The resolved part of the force $\sigma_x \sin \theta$ parallel to the plane is $\sigma_x \sin\theta \cos\theta$ so, resolving parallel to plane gives

$$\tau \times 1 = \sigma_x \sin \theta \times \cos \theta$$

$$\begin{array}{ccc}\text{(force)} & \text{(force)} & \text{(resolved parallel to plane)}\end{array}$$

$$\tau = \sigma_x \sin \theta \cos \theta = \tfrac{1}{2} \sigma_x \sin 2\theta \tag{4.4}$$

For τ to be a maximum $\dfrac{d\tau}{d\theta} = 0$

or $\dfrac{d}{d\theta}\left(\dfrac{\sigma_x \sin 2\theta}{2}\right) = 0$

$$\sigma_x \cos 2\theta = 0$$
$$\therefore \cos 2\theta = 0$$
$$2\theta = 90°$$
$$\theta = 45°$$

Substituting back into equation (4.4) gives:

$$\tau_{max} = \sigma_x \times \frac{1}{\sqrt{2}} \times \frac{1}{\sqrt{2}}$$

$$= \frac{\sigma_x}{2} \tag{4.5}$$

So, the maximum value of shear stress in a bar subjected to an applied direct tensile stress occurs on a plane at 45° to the plane carrying the applied stress, and has a value of half the applied stress.

This result accounts for the fact that a tensile test taken to failure often exhibits a fracture at approximately 45° to the axis of applied tension. The standard 'cup and cone' and brittle type fractures indicate this tendency, Fig. 4.5, showing that the shear stress has been significant in the final cause of failure.

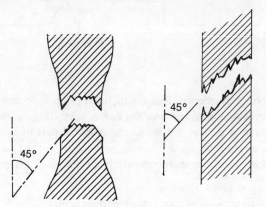

Fig. 4.5

4.3   Two-dimensional direct stress systems

A more complex stress situation arises when an additional applied stress has a component which is perpendicular to the direction of the original stress. For example in Fig. 4.6, the rectangular solid shown is subjected to a direct tensile stress σ_x in the x-direction, and a direct tensile stress σ_y in the y-direction.

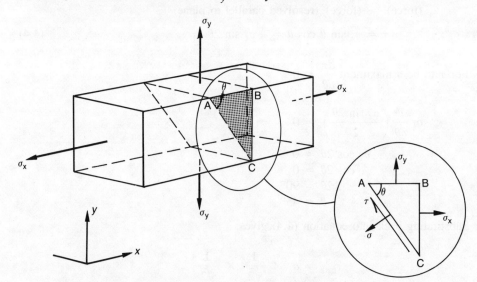

Fig. 4.6

To find the stresses that act on a plane inclined at an angle θ to the x-direction, consider a small triangular element ABC of the solid which includes the plane in question. Let σ and τ be the direct and shear stress that act on the plane as a result of the applied stresses σ_x and σ_y. Again, for convenience, we will assume that the area of the plane AC is unity. So that the area of BC is $\sin \theta$ and the area of AB is $\cos \theta$. The *force* on BC is thus $\sigma_x \sin \theta$, and the *force* on AB is $\sigma_y \cos \theta$.

For equilibrium of *forces* acting on the triangular element, resolving perpendicular to the plane AC

$$\sigma \times 1 = \sigma_x \sin \theta \sin \theta + \sigma_y \cos \theta \cos \theta$$

(stress × area   (stress × area ×
× resolved part)   resolved part)

$$\sigma = \sigma_x \sin^2 \theta + \sigma_y \cos^2 \theta \tag{4.6}$$

Resolving parallel to plane AC

$$\tau = \sigma_x \sin \theta \cos \theta - \sigma_y \cos \theta \sin \theta$$

(stress × area ×   (stress × area ×
resolved part)   resolved part)

$$= (\sigma_x - \sigma_y) \sin \theta \cos \theta$$

$$= \left(\frac{\sigma_x - \sigma_y}{2}\right) \sin 2\theta \tag{4.7}$$

Equations (4.6) and (4.7) show that it is a simple matter to compute the direct and shear stresses that act on any given plane.

Example 4.4

At a point in a material a direct tensile stress of 80 MN/m^2 acts perpendicularly to a direct compressive stress of 60 MN/m^2. Calculate the value of direct and shear stresses, and the resultant stress, that acts on a plane inclined at an angle of 60° to the plane carrying the 80 MN/m^2 stress.

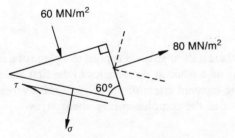

For the equilibrium of *forces* acting on a triangular element whose inclined face is considered to be of unit area, resolving perpendicularly to the plane:

$$\sigma = 80 \cos 60° \cos 60° - 60 \sin 60° \sin 60°$$

$$= 80 \times \tfrac{1}{2} \times \tfrac{1}{2} - 60 \times \frac{\sqrt{3}}{2} \times \frac{\sqrt{3}}{2}$$

$$= 20 - 45$$

$$= -25 \text{ MN/m}^2$$

and resolving parallel to plane

$$= 80 \cos 60 \sin 60 + 60 \sin 60 \cos 60$$
$$= 140 \cos 60 \sin 60$$

$$= 140 \times \frac{1}{2} \times \frac{\sqrt{3}}{2}$$

$$= 60.62 \text{ MN/m}^2$$

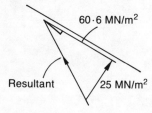

Thus a compressive stress of 25 MN/m^2 and a shear stress of 60·6 MN/m^2 act on the plane. The resulting stress is given by the parallelogram law and Pythagoras, thus:

$$\text{resultant stress} = \sqrt{60 \cdot 6^2 + 25^2}$$
$$= \underline{65 \cdot 55} \text{ MN/m}^2$$

at an angle $\psi = \tan^{-1} \dfrac{25}{60 \cdot 6} = \underline{22° \ 25'}$ to the plane

4.4   The complementary shear stress

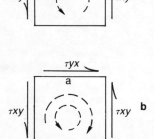

Fig. 4.7

A similar treatment is used when the additional stress is a shear stress, but here there is a major factor which must always be taken into account when considering shear stresses.

Consider the cube of material of side 'a' shown in Fig. 4.7a to which there is being applied a shear stress τ_{xy}. It is clear that such an arrangement is not an equilibrium case since the cube is subjected to a couple of magnitude $\tau_{xy} \times$ a which would cause the cube to rotate. Clearly an opposing couple of equal value is required to maintain an equilibrium situation. Let this opposing couple be supplied by a shear stress τ_{xy} as shown in Fig. 4.7b.

For equilibrium of moments it is evident that:

$$\tau_{xy}.\text{a} = \tau_{yx}.\text{a}$$
or $$\tau_{xy} = \tau_{yx}$$

This result means that in all cases where a shear stress is applied to a plane of a material there is set up an equal shear stress on a plane at right-angles to the first and in such a direction as to oppose the turning moment effect of the original shear stress. This compensating shear stress is known as the **complementary shear stress**.

4.5   Two-dimensional shear stress systems

Suppose that a shear stress τ_{xy} was applied to the parallel faces of a rectangular solid. A complementary shear stress of equal value would automatically result on the adjacent faces, (Fig. 4.8). Note that the directions of the arrows representing the shear stresses are always towards or away from the corners of the solid.

To find the stresses that act on a plane inclined at an angle θ to the x-direction, consider a triangular element ABC which includes the plane under consideration. If σ and τ are the direct and shear stresses that act on the plane then for equilibrium of the *forces* that act on the element, again assuming the area of the plane AC to be unity, resolving parallel to the plane:

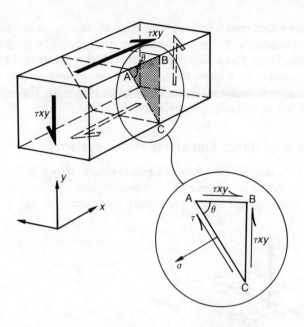

Fig. 4.8

$$\tau \times 1 = \tau_{xy}. \cos \theta. \cos \theta - \tau_{xy}. \sin \theta. \sin \theta$$
$$\text{(stress} \times \text{area} \times \qquad \text{(stress} \times \text{area} \times$$
$$\text{resolved part)} \qquad \text{resolved part)}$$

which reduces to:

$$\tau = \tau_{xy} (\cos^2 \theta - \sin^2 \theta)$$
or $$\tau = \tau_{xy} \cos 2\theta \qquad\qquad (4.8)$$

Resolving perpendicularly to plane AC

$$\sigma \times 1 = \tau_{xy} \cos \theta. \sin \theta + \tau_{xy} \sin \theta. \cos \theta$$
$$\text{(stress} \times \text{area} \times \ + \text{(stress} \times \text{area} \times$$
$$\text{resolving part)} \qquad \text{resolved part)}$$

which reduces to:

$$\sigma = \tau_{xy} \sin \theta. \cos \theta$$

or $$\sigma = \tau_{xy} \frac{\sin 2\theta}{2} \qquad\qquad (4.9)$$

From equation (4.9), the direct stress σ will be a maximum or a minimum when

$$\frac{d\sigma}{d\theta} = 0$$

i.e. when $\tau_{xy} \cos 2\theta = 0$
or $\cos 2\theta = 0$
$$2\theta = 90°, 270°$$
$$\theta = 45°, 135°$$

Substituting these values of θ into equation (4.8) gives $\tau = 0$.

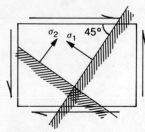

Fig. 4.9

Hence the direct stress which results from applied shear stresses τ_{xy}, is a maximum or a minimum on planes inclined at 45° to the planes carrying τ_{xy}, and on these planes the shear stress τ is zero. These planes in a material are called the **principal planes**. One of the principal stresses, σ_1 will be the maximum direct stress acting in the material, and the other, σ_2, will be the minimum direct stress (Fig. 4.9). The two principal stresses therefore act in mutually perpendicular directions.

4.6   Two-dimensional direct and shear stress systems

The general case of a two-dimensional complex stress system is shown in Fig. 4.10. Here, direct stresses σ_x and σ_y act on planes in mutually perpendicular directions, together with a shear stress τ_{xy} and its complementary shear stress.

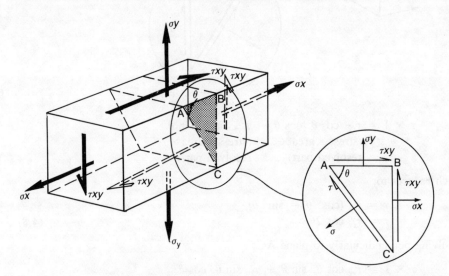

Fig. 4.10

As before, considering the equilibrium of *forces* acting on a triangular element ABC whose face AC we assume to be of unit area, resolving parallel to plane AC

$$\tau \times 1 = \sigma_x \sin \theta . \cos \theta - \sigma_y \cos \theta \sin \theta + \tau_{xy} \cos \theta \cos \theta$$
$$- \tau_{xy} \sin \theta . \sin \theta$$

$$\therefore \tau = (\sigma_x - \sigma_y) \sin \theta . \cos \theta + \tau_{xy} (\cos^2\theta - \sin^2\theta)$$

$$\therefore \tau = \left(\frac{\sigma_x - \sigma_y}{2}\right) \sin 2\theta + \tau_{xy} \cos 2\theta \qquad (4.10)$$

and, resolving perpendicularly to plane AC

$$\sigma = \sigma_x \sin \theta . \sin \theta + \sigma_y \cos \theta . \cos \theta$$
$$+ \tau_{xy} \sin \theta . \cos \theta + \tau_{xy} \cos \theta . \sin \theta$$

$$\therefore \sigma = \sigma_x \sin^2\theta + \sigma_y \cos^2\theta + \tau_{xy} \sin 2\theta \qquad (4.11)$$

Using equations 4.11 σ is a maximum or minimum when $\dfrac{d\sigma}{d\theta} = 0$

i.e. when $\dfrac{d}{d\theta}(\sigma_x \sin^2\theta + \sigma_y \cos^2\theta + \tau_{xy} \sin 2\theta) = 0$

i.e.
$$\sigma_x\, 2\sin\theta\, \cos\theta - \sigma_y\, 2\cos\theta\, \sin\theta + 2\tau_{xy} \cos 2\theta = 0$$
$$(\sigma_x - \sigma_y) \sin 2\theta + 2\tau_{xy} \cos 2\theta = 0$$

Dividing through by $\cos 2\theta$ and re-arranging

$$(\sigma_x - \sigma_y) \tan 2\theta = -2\tau_{xy}$$

or $\tan 2\theta = -\dfrac{2\tau_{xy}}{\sigma_x - \sigma_y}$

$$2\theta = \tan^{-1}\left(\frac{-2\tau_{xy}}{\sigma_x - \sigma_y}\right) \text{ or } \tan^{-1}\left(\frac{-2\tau_{xy}}{\sigma_x - \sigma_y}\right) + 180°$$

So that
$$\theta = \tfrac{1}{2}\tan^{-1}\left(\frac{-2\tau_{xy}}{\sigma_x - \sigma_y}\right) \text{ or } \tfrac{1}{2}\tan^{-1}\left(\frac{-2\tau_{xy}}{\sigma_x - \sigma_y}\right) + 90° \qquad (4.12)$$

Again this gives the result that σ has maximum and minimum values on two planes separated by 90°, the Principal Planes.

From equation (4.10)

$$\tau = 0 \quad \text{when} \quad \left(\frac{\sigma_x - \sigma_y}{2}\right) \sin 2\theta = -\tau_{xy} \cos 2\theta$$

re-arranging, $\tan 2\theta = \dfrac{-2\tau_{xy}}{\sigma_x - \sigma_y}$

whence $\theta = \tfrac{1}{2}\tan^{-1}\left(\dfrac{-2\tau_{xy}}{\sigma_x - \sigma_y}\right) \quad \text{or} \quad \tfrac{1}{2}\tan^{-1}\left(\dfrac{-2\tau_{xy}}{\sigma_x - \sigma_y}\right) + 90°$

This angle is the same as the angle of the principal planes, and so we deduce that the principal planes in a material are mutually perpendicular; they carry no shear stress; one of the planes carries the maximum value of direct stress σ_1 (the maximum principal stress); the other carries the minimum value of direct stress σ_2, (the minimum principal stress).

Example 4.5

At a point in a material the direct stresses are 40 MN/m² (tensile) and 60 MN/m² (tensile) in mutually perpendicular directions. On the planes carrying the applied direct stresses, there also acts an applied shear stress of value 30 MN/m². Calculate, relative to the direction of the 40 MN/m² stress, the angles of the principal planes, and the values of the principal stresses.

A complementary shear stress of 30 MN/m² acts on a plane at right-angles to the plane carrying the applied shear stress.

For the equilibrium of a triangular element at a point in the material, let σ be a principal stress and θ the angle that the principal plane makes with the direction of the 40 MN/m² stress. *There is no shear stress on this principal plane.*

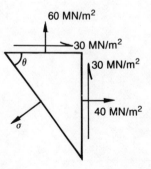

Resolving *forces* perpendicularly to the plane, and assuming the plane to be of unit area

$$\sigma \times 1 = 40 \sin\theta . \sin\theta + 60 \cos\theta . \cos\theta + 30 \sin\theta . \cos\theta$$
$$\text{(force)} \quad \text{(resolved part)}$$
$$+ 30 \cos\theta . \sin\theta$$

$$\therefore \; \sigma = 40 \sin^2\theta + 60 \cos^2\theta + 15 \sin2\theta$$

Resolving *forces* parallel to the plane

$$0 = 40 \sin\theta . \cos\theta - 60 \cos\theta . \sin\theta + 30 . \cos\theta \cos\theta$$
$$\text{(force)} \quad \text{(resolved part)}$$
$$- 30 \sin\theta . \sin\theta$$

$$= 40 \sin\theta \cos\theta - 60 . \sin\theta \cos\theta + 30 (\cos^2\theta - \sin^2\theta)$$
$$= -20 \sin\theta \cos\theta + 30 (\cos^2\theta - \sin^2\theta)$$
$$= -10 \sin 2\theta + 30 \cos 2\theta$$

Re-arranging

$$\frac{\sin 2\theta}{\cos 2\theta} = \frac{30}{10}$$

or $\qquad \tan 2\theta = 3$

whence $\qquad 2\theta = 71° \; 34'$ or $251° \; 34'$
$$\theta = 35° \; 47' \quad \text{or} \quad 125° \; 47'$$

Thus there are two principal planes, 90° apart which carry no shear stress. The values of the principal stresses which act on these planes can be found by substituting the angles into the equation for σ. Thus

$$\sigma_1 = 40 \sin^2(35° \; 47') + 60 \cos^2(35° \; 47') + 15 \sin(71° \; 34')$$
$$= 13 \cdot 676 + 39 \cdot 486 + 14 \cdot 23$$
$$= \underline{67 \cdot 4 \; MN/m^2}$$

$$\sigma_2 = 40 \sin^2(125° \; 47') + 60 \cos^2(125° \; 47') + 15 \sin(251° \; 34')$$
$$= 26 \cdot 324 + 20 \cdot 514 - 14 \cdot 23$$
$$= \underline{32 \cdot 6 \; MN/m^2}$$

The maximum principal stress σ_1 is $67 \cdot 4 \; MN/m^2$ on a plane making an angle of 35° 47′ to the direction of the 40 MN/m^2 stress, and the minimum principal stress is $32 \cdot 6 \; MN/m^2$ on a plane making an angle of 125° 47′ to the direction of the 40 MN/m^2 stress.

4.7   The maximum shearing stress

The principal planes define the directions of zero shearing stress. It is evident then that on some intermediate plane the shear stress will reach a maximum value.

From equation (4.10) we have:

$$\tau = \frac{\sigma_x - \sigma_y}{2} \sin 2\theta + \tau_{xy} \cos 2\theta$$

Differentiating $\dfrac{d\tau}{d\theta} = \left(\dfrac{\sigma_x - \sigma_y}{2}\right) 2 \cos 2\theta - \tau_{xy} 2 \sin 2\theta$

$$= 0 \text{ for a maximum}$$

i.e. $(\sigma_x - \sigma_y) \cos 2\theta = 2 \tau_{xy} \sin 2\theta$

or $\tan 2\theta = \left(\dfrac{\sigma_x - \sigma_y}{2\tau_{xy}}\right)$

inverting $\cos 2\theta = \dfrac{2\tau_{xy}}{\sigma_x - \sigma_y}$

whence $\theta = \frac{1}{2} \cos^{-1}\left(\dfrac{2\tau_{xy}}{\sigma_x - \sigma_y}\right)$ (4.13)

Comparing this result with equation (4.12), and using the trigonometrical identity $\cos A = -\tan (A+90°)$ it is clear that the planes of maximum shear stress are inclined at 45° to the principal planes (Fig. 4.11).

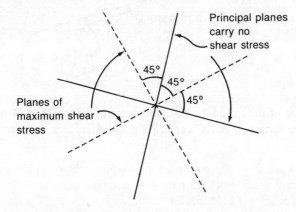

Fig. 4.11

It can be shown that the value of the maximum shear stress is half the difference between the principal stresses

i.e. $\tau\text{max} = \dfrac{\sigma_1 - \sigma_2}{2}$ (4.14)

Example 4.6

A thin cyclinder with closed ends 75 mm diameter and wall thickness 5 mm is subjected to an internal pressure of 60 bar, and also a torque of 1500 Nm. Determine the maximum and minimum principal stresses and the maximum shearing stress in the cylinder.

The internal pressure will give rise to hoop and axial stresses where;

$$\sigma_h = \frac{pd}{2t} = \frac{60 \times 10^5 \times 0{\cdot}075}{2 \times 0{\cdot}005} = 45 \text{ MN/m}^2$$

and $\qquad \sigma_a = \dfrac{pd}{4t} = \dfrac{45}{2} = 22 \cdot 5 \text{ MN/m}^2$

The applied torque will produce a shear stress τ where:

$$\tau = \frac{Tr}{J}$$

from the torsion formula, and T is the applied torque, r the radius of the shaft and J the polar second moment of area

$$= \frac{1500 \times 0 \cdot 075/2}{\pi(0 \cdot 0754^4 - 0 \cdot 065^4)/32}$$

$$= 41 \cdot 55 \text{ MN/m}^2$$

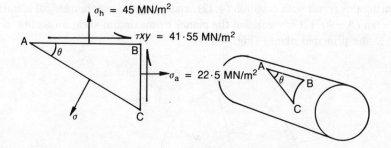

At a point on the surface of the cylinder, these stresses will act together. If ABC is a triangular element of the surface, with AC a principal plane of unit area, then, resolving parallel to the plane AC

$$0 = 22 \cdot 5 \sin \theta \cos \theta - 45 \cos \theta \sin \theta + 41 \cdot 55 \cos^2\theta - 41 \cdot 55 \sin^2\theta$$
$$= -22 \cdot 5 \sin \theta \cos \theta + 41 \cdot 55 (\cos^2\theta - \sin^2\theta)$$
$$= -11 \cdot 25 \sin 2\theta + 41 \cdot 55 \cos 2\theta$$

Re-arranging,

$$\frac{\sin 2\theta}{\cos 2\theta} = \tan 2\theta = \frac{41 \cdot 55}{11 \cdot 25} = 3 \cdot 693$$

whence $\qquad 2\theta = 74° \; 51', \; 254° \; 51'$
and $\qquad \theta = 37° \; 25', \; 127° \; 25'$

Resolving perpendicularly to plane AC gives

$$\sigma = 22 \cdot 5 \sin \theta \sin \theta + 45 \cos \theta \cos \theta + 41 \cdot 55 \sin \theta \cos \theta$$
$$\qquad + 41 \cdot 55 \cos \theta \sin \theta$$
$$\sigma = 22 \cdot 5 \sin^2\theta + 45 \cos^2\theta + 41 \cdot 55 \sin 2\theta$$

when $\qquad \theta = 37° \; 25'$
$$\sigma = 22 \cdot 5 \sin^2(37° \; 25') + 45 \cos^2(37° \; 25') + 41 \cdot 55 \sin(74° \; 51')$$
$$= 8 \cdot 307 + 28 \cdot 387 + 40 \cdot 106$$
$$= 76 \cdot 8 \text{ MN/m}^2$$

and when $\theta = 127° \ 25'$
$$\sigma = 22 \cdot 5 \sin^2(127° \ 25') + 45 \cos^2(127° \ 25')$$
$$+41 \cdot 55 \sin(245° \ 51')$$
$$= 14 \cdot 193 + 16 \cdot 613 - 40 \cdot 106$$
$$= - \ 9 \cdot 3 \ \text{MN/m}^2$$

Hence the maximum principal stress, $\sigma_1 = 76 \cdot 8 \ \text{MN/m}^2$ (tensile) and the minimum principal stress, $\sigma_2 = 9 \cdot 3 \ \text{MN/m}^2$ (compressive). The maximum shear stress, τ_{max} is given by:

$$\tau_{max} = \frac{\sigma_1 - \sigma_2}{2}$$

$$= \frac{76 \cdot 8 \ - \ (-9 \cdot 3)}{2}$$

$$= 43 \cdot 05 \ \text{N/m}^2$$

Guided solution

The shank of a bolt is reduced so that it is equal to the smaller diameter of the thread.

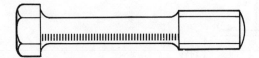

When being tightened a torque is supplied by a spanner and the bolt also experiences a tensile stress due to the axial load. If an applied torque of 40 Nm produces a shear stress of 120 MN/m^2 and a direct stress of 100 MN/m^2, calculate the maximum torque that can be applied if the principal stress must not exceed 280 MN/m^2.

Step 1. Draw a triangular element of the shank of the bolt and enter on it
  a the direct stress
  b the shear stress and its complementary shear stress
  c a principal stress on the inclined plane at an angle θ to the axial direction.

Step 2. By resolving *forces* parallel and perpendicular to the principal plane write down two equations of equilibrium.

Step 3. Since there is no shear stress on a principal plane the equation obtained by resolving parallel to the plane will yield the angle of the plane θ(33° 42′ and 123° 42′).

Step 4. Enter the values of θ into the equation for σ to find the maximum principal stress (180 MN/m^2).

Step 5. The limiting principal stress is 280 MN/m^2 and so the applied torque can be increased in the ratio of 280 : 180. Calculate the new torque (62·2 Nm).

Further practice can now be obtained by attempting the first four questions of Problems 4.2.

4.8   Mohr's circle of stress

Equations (4.10) and (4.11) for τ and σ show that the values of direct and shear stress on a plane inclined at an angle vary continually with the value of θ. Eliminating θ between the two equations and re-arranging gives the result:

$$[\sigma - \tfrac{1}{2}(\sigma_x + \sigma_y)]^2 + \tau^2 = [\tfrac{1}{2}(\sigma_x - \sigma_y)]^2 + [\tau_{xy}]^2 \qquad (4.15)$$

This equation is of the form

$$(x - a)^2 + y^2 = r^2$$

which is the equation of a circle of radius r with its centre at the point (a, o).

Equation 4.15 shows then that the values of σ and τ all lie on a circle of radius $\sqrt{[\tfrac{1}{2}(\sigma_x - \sigma_y)]^2 + [\tau_{xy}]^2}$ with its centre at the point $(\tfrac{1}{2}[\sigma_x + \sigma_y], 0)$. This circle, which defines all the possible values of σ and τ is known as **Mohr's circle of stress**. The construction of Mohr's circle very much simplifies the solution of complex stress problems, and proceeds as follows. For the most general case of complex stress, we would normally be given two direct stresses σ_x and σ_y, and a shear stress and its accompanying complementary shear stress τ_{xy}.

Let σ and τ be the direct and shear stresses that result on a plane inclined at an angle θ to the x-direction, (Fig. 4.12).

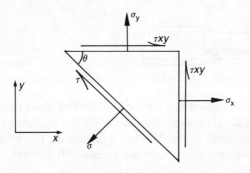

Fig. 4.12

Step 1.   Draw rectangular axes for σ and τ using the same scales for each.

Step 2.   Plot the values of the direct stresses σ_x and σ_y on the axis, remembering that a tensile stress is positive and a compressive stress negative.

Step 3.   Plot ordinates to represent $\pm\tau_{xy}$ using the sign convention that shear stresses tending to produce a clockwise couple are positive, i.e. $+\tau_{xy}$ to correspond with σ_x and $-\tau_{xy}$ to correspond with σ_y.

Step 4.   Join the points, (σ_x, τ_{xy}) and $(\sigma_y, -\tau_{xy})$. This line will form a diameter of Mohr's circle.

Step 5.   Draw the circle using the intersection of the line and the σ-axis as the centre.

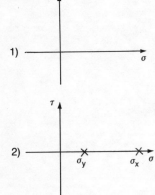

Mohr's circle of stress, once drawn, can be used to find the direct and shear stresses on any plane, the principal stresses, and the maximum shear stress and the planes on which they act. For example, consider the Mohr's circle in Fig. 4.13 and also referring to Fig. 4.12.

(a) The radius OP represents the plane on which σ_x and τ_{xy} act, and the radius OQ represents the plane on which σ_y and τ_{xy} act.

(b) These radii are 180° apart, and so we deduce that planes which are actually 90° apart, appear on Mohr's circle 180° apart. This can be extended so that any actual angle θ appears as 2θ on Mohr's circle.

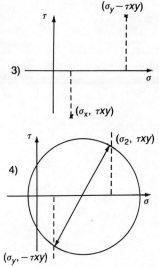

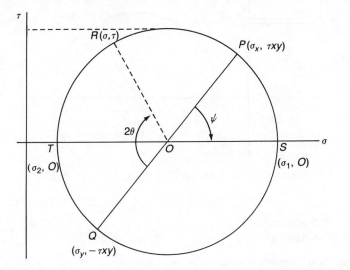

Fig. 4.13

(c) σ and τ act on a plane which is *measured clockwise* to the plane carrying σ_y. Thus on Mohr's circle we measure $2\theta°$ *clockwise* from OQ, the plane carrying σ_y, to give the point R.

(d) The maximum principal stress, σ_1, lies at S and the minimum principal stress, σ_2, lies at T.

(e) The principal plane carrying the maximum principal stress is OS, an angle $\psi°$ from OP measured clockwise. In Fig. 4.15 then the principal plane is $\psi°/2$ from the plane carrying σ_x, again measured clockwise.

(f) The principal plane carrying the minimum principal stress is OT 180° (clockwise) from OS on Mohr's circle, but 90° (clockwise) from the other principal plane in Fig. 4.12.

(g) The maximum shear stress is the radius of the circle $= \dfrac{\sigma_1 - \sigma_2}{2}$.

Example 4.7

At a point in a material the two-dimensional stresses forming a complex stress system are:

$$\sigma_x = 75 \text{ MN/m}^2 \text{ (tensile)}$$
$$\sigma_y = 45 \text{ MN/m}^2 \text{ (tensile)}$$
$$\tau_{xy} = 30 \text{ MN/m}^2$$

Find the values of the principal stresses and the maximum shear stress, and the angle of the planes on which they act.

On $\sigma - \tau$ axes a circle is constructed with a line joining (75,30) and (45, −30) as diameter. The intercepts of the circle on the σ-axis are:

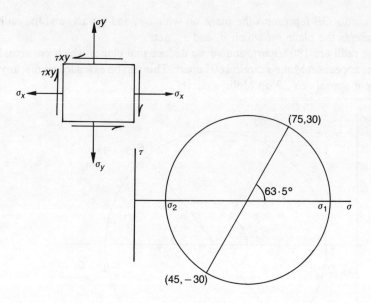

$$\sigma_1 = 93 \cdot 5 \ \text{MN/m}^2$$
$$\sigma_2 = 26 \cdot 5 \ \text{MN/m}^2$$

The maximum shear stress is the radius of the circle

$$= \frac{\sigma_1 - \sigma_2}{2}$$

$$= \frac{93 \cdot 5 - 26 \cdot 5}{2} = 33 \cdot 5 \ \text{MN/m}^2$$

The angle that the principal plane carrying σ_1 makes with the plane carrying σ_x is $63 \cdot 5°/2$

$$= \underline{31° \ 45'} \ \text{clockwise}$$

Example 4.8

At a point in a material a shear stress of 35 MN/m^2 (clockwise) acts together with a tensile stress of 60 MN/m^2. If the maximum allowable direct stress in the material is 80 MN/m^2, calculate the value of direct stress that may be applied at right-angles to the 60 MN/m^2 stress. Since the maximum direct stress must not exceed 80 MN/m^2, then this is the value that σ_1 must have.

Thus we have two co-ordinate points that we can plot on $\sigma - \tau$ axes, viz. P (60,35) and Q (80,0).

To find the centre of the circle, construct the perpendicular bisector of the line joining the two known points. Where this line intersects the σ_x-axis is the centre of the circle, O. Using OQ as the radius the circle is drawn.

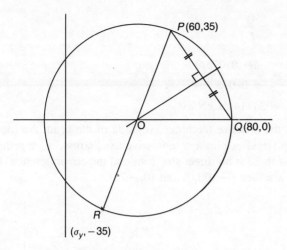

The line OP represents the plane carrying the 60 MN/m² direct stress. The plane carrying the unknown direct stress is 180° from this on the Mohr's circle. This plane is thus represented by OR. The σ co-ordinate of the point R is the value of the unknown direct stress. Thus from the diagram, $\sigma_y = 16 \cdot 5$ MN/m².

N.B. The τ co-ordinate of R is -35 MN/m², i.e. the anti-clockwise shear stress.

Example 4.9

A hollow drive shaft 75 mm external diameter and 50 mm internal diameter is subjected to an axial thrust of 40 kN when transmitting 500 kW at 800 rev/min. If the greater principal stress is not to exceed 100 MN/m², calculate the greatest value of bending moment that can be applied to the shaft.

$$\text{Axial stress, } \sigma_a = \frac{\text{Axial thrust}}{\text{Area of cross-section}}$$

$$= \frac{40 \times 10^3}{\pi(0 \cdot 075^2 - 0 \cdot 05^2)/4}$$

$$= \underline{16 \cdot 3 \text{ MN/m}^2}$$

$$\text{Applied torque} = \frac{\text{power} \times 60 \times 10^3}{2\pi \times \text{speed}} = \frac{500 \times 60 \times 10^3}{2\pi \times 800}$$

$$= \underline{5 \cdot 97 \text{ kNm}}$$

$$\text{From } \frac{\tau}{r} = \frac{T}{J}, \tau = \frac{Tr}{J} = \frac{5 \cdot 97 \times 10^3 \times \dfrac{0 \cdot 075}{2}}{\pi \, 0 \cdot 075^4/32}$$

$$= \underline{72 \cdot 07 \text{ MN/m}^2}$$

The bending stress, σ_b

$$= \frac{My}{I}$$

$$= \frac{M \cdot . 0 \cdot 075/2}{\pi \, 0 \cdot 075^4/64}$$

$$= 24 \cdot 14 \, M \, kN/m^2$$

The stresses acting on the concave (compressive) side of the beam are then:

A compressive bending stress σ_b plus a compressive axial stress, σ_a, together with a shear stress τ. Note that there is no direct stress around the circumference. Known points on the $\sigma - \tau$ axes are then $(-100,0)$ and $(0,-72)$.

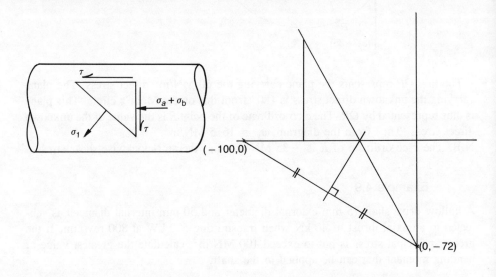

The perpendicular bisector of the line joining these two points will cut the σ-axis at the centre of the circle, O. From the diagram the co-ordinates of other end of the diameter through O and the point $(0, -72)$ are $(50,72)$.

The *total* stress on the lateral plane is then 50 MN/m².

Thus $\sigma_a + \sigma_b = 50 \times 10^6 \, N/m^2$

or $16 \cdot 3 \times 10^6 + 24 \cdot 14 \times 10^3 \times M = 50 \times 10^6$

whence $M = 1 \cdot 396$ kNm.

Guided solution

A thin cylinder, 200 mm diameter and with a wall thickness of 3 mm is subjected to an internal pressure of 40 bar, and an axial tensile load of 70 kN. Calculate the direct and shear stresses that result on a plane inclined at 30° to the axis of the cylinder.

Step 1. Using the formulae for the hoop stress and axial stress for a thin cylinder calculate σ_h and σ_a ($\sigma_h = 133$ MN/m², $\sigma_a = 66 \cdot 5$ MN/m²).

Step 2. Calculate the extra axial stress due to the applied axial load, ($37 \cdot 14$ MN/m²).

Step 3. Find the total axial stress by adding the result of step 2 to the value of a in step 1 ($103 \cdot 64$ MN/m²).

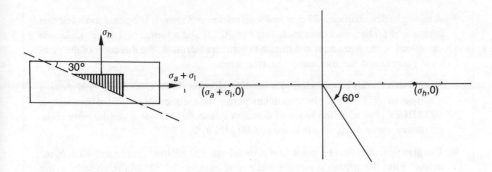

Step 4. Draw a triangular element of the cylinder's wall, and show on it the direct stresses that act. Are there any shear stresses?

Step 5. Plot, on $\sigma - \tau$ axes, the direct stresses co-ordinates, and draw Mohr's circle of stress.

Step 6. Measure $2 \times 30° = 60°$ clockwise from the plane carrying σ_h, and note the co-ordinates of the intersection of the circle to give the required values. $(124 \cdot 5 \text{ MN/m}^2, 12 \cdot 25 \text{ MN/m}^2)$.

Problems 4.2

1 At a point in a material perpendicular direct stresses of 80 MN/m^2 and 50 MN/m^2, both tensile act together. Determine the direct and shear stresses that result on a plane inclined at 30° to the axis of the 50 MN/m^2 stress.

2 A thin cyclinder 100 mm diameter and with a wall thickness of 5 mm is subjected to an internal pressure of 50 bar and a torque of 2 kNm. Calculate the direct and shear stresses that act on a plane inclined at 20° to the axis of the cylinder.

3 A round bar of diameter 50 mm is subjected to a bending moment of 75 kNm and a torque of 90 kNm. Calculate the maximum direct and shearing stresses in the bar.

4 At a point in a material there acts a shear stress of 55 MN/m^2. On the planes on which the shear stress is conventionally positive there acts a tensile stress of 70 MN/m^2, and on a perpendicular plane there acts a compressive stress of 25 MN/m^2. Calculate
a the values of the principal stresses
b the angles of the principal planes relative to the plane carrying the 70 MN/m^2 stress
c the value of the maximum shear stress.

5 The principal stresses at a point in a material are 80 MN/m^2 tension and 30 MN/m^2 tension. Calculate the values of direct and shear stresses acting on a plane inclined at 30° to the plane carrying the larger principal stress, and the magnitude of the resultant stress on this plane.

6 At a point in a material there acts a tensile stress of 60 MN/m^2 and a compressive stress of 40 MN/m^2 on planes which are mutually perpendicular. If the maximum direct stress in the material is limited to 80 MN/m^2 find the limiting value shear stress that can be applied to the planes. What then will be the maximum value of shear stress that acts in the material?

7 A thin cylinder, diameter 75 mm and wall thickness 3 mm, is subjected to an internal pressure of 60 bar, an axial tensile force of 20 kN and a torque of 1 kNm. Calculate the values of the maximum and minimum principal stresses, the direction of the principal planes, and the maximum shearing stress.

8 A solid drive shaft is subjected to a bending moment of 25 kNm whilst transmitting a torque of 40 kNm. If the maximum permissible direct stress in the material is 100 MN/m^2, calculate the required diameter of the shaft. What is the diameter if the maximum shear stress is not to exceed 60 MN/m^2?

9 The principal stresses at a point in a material are 120 MN/m^2 tensile and 40 MN/m^2 tensile. Find the *strain* on a plane inclined at an angle of 30° to the plane carrying the greater principal stress. $E = 200$ GN/m^2

Practical exercise

Attach two electrical resistance strain gauges mounted at right-angles to either:

a the outer surface of a cylindrical pressure vessel so that one gauge lies along the axis, and the other around the circumference;
or **b** the upper surface of a thin flat horizontal plate which is being subjected to tensile stresses in mutually perpendicular directions.

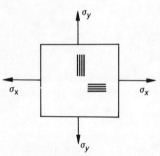

Using acceptable **values** for the elastic constants for the material used for the components (e.g. for mild steel $E = 208$ GN/m^2, $\nu = 0\cdot3$) calculate the stresses in the direction of the gauges.

Draw Mohr's circle for the given stresses and compare with a circle drawn from purely calculated values of the stresses. Determine the values of direct and shear stresses that act on some plane (say 30°) to the axial direction and compare with calculated results.

Comment on the experimental method, the validity of the results and the assumptions that the theory and practice make.

5 Combined bending and direct stress

On completion of this chapter you will be able to solve a range of problems in which the forces applied to a structure cause a combination of bending and direct stress.

5.1 second moment of area

In Section 2.8 and Appendix 2 the concept of second moment of area (I) was introduced, in particular the second moments of area of particular cross-sectional shapes, viz.

The second moment of area of a rectangle about one edge, $I_{XX} = \dfrac{bd^3}{3}$

The Second moment of area of a rectangle about neutral axis, $I_{NA} = \dfrac{bd^3}{12}$

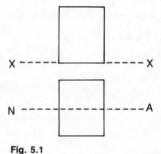

Fig. 5.1

In addition it is possible to find the second moment of area of a rectangle about *any* axis which is parallel to the axis through the neutral axis using the **parallel axis theorem**.

Thus $I_{XX} = I_{NA} + Ah^2$

Where A is the area of the rectangle and h the distance between the parallel axes. This equation is particularly useful when finding the second moment of area of a shape composed of simple rectangles, for example the simplified cross-section of 'T' and 'I' section beams (Fig. 5.2).

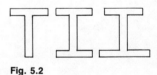

Fig. 5.2

Example 5.1

Calculate the value of the second moment of area about the neutral axis of the stylised 'T' section beam shown, (Fig. 5.3).

We first need to find the position of the neutral axis. Let this be a distance $\bar{y}$ from the base XX of the 'T'.

Using the principle of the *First* moment of area the sum of the first moments of area of each individual shape about XX is equal to the first moment of area of the total about XX.

Thus if we divide the total shape into two individual rectangles A and B:

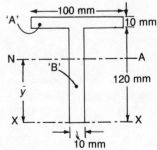

Fig. 5.3

First moment of area of A about XX + First moment of area of B about XX
= First moment of area or total about XX

or   Area of A × distance of centroid of A from XX

+ Area of B × distance of centroid of B from XX
= Area of total × distance of centroid of total from XX

or $(100 \times 10) \times 125 + (120 \times 10) \times 60 = (100 \times 10 + 120 \times 10) \times \bar{y}$

$125\ 000 + 72\ 000 = 2200\ \bar{y}$

$$\therefore \bar{y} = \frac{197\ 000}{2200} = \underline{89 \cdot 545}\ \text{mm}.$$

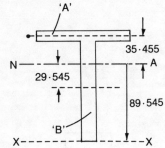

Fig. 5.4

Comparing this result with the diagram (Fig. 5.3) the answer looks reasonable enough. To find the second moment of area about the neutral axis, i.e. an axis parallel to, and 89·545 mm from, XX, there are two alternatives.

Method 1, in which the second moments of area of the individual rectangles A and B about NA are found and added (Fig. 5.4).

Thus, second moment of area of rectangle A about NA using the parallel axis theorem is given by

$$I_{NA} = I_{na} + Ah^2$$

$$= \frac{bd^3}{12} + b.d.h^2$$

$$= \frac{100 \times 10^3}{12} + (100 \times 10) \times 35 \cdot 455^2$$

$$= 8333 \cdot 333 + 1\ 257\ 057$$
$$= 1265\ 390\ \text{mm}^4$$

Similarly I_{NA} for rectangle B

$$= \frac{bd^3}{12} + Ah^2$$

$$= \frac{10 \times 120^2}{12} + (10 \times 120) \times 29 \cdot 545^2$$

$$= 1\ 440\ 000 + 1\ 047\ 488 \cdot 4$$
$$= 2\ 487\ 488 \cdot 4\ \text{mm}^4$$

Total I_{NA} for whole 'T' section $= 1\ 265\ 390 + 2\ 487\ 488$
$$= 3\ 752\ 878\ \text{mm}^4$$
$$= \underline{3 \cdot 753 \times 10^{-6}\ \text{m}^4}$$

Method 2, in which the 'T' shape is composed of a rectangle 100×130 mm *less* two rectangles each 45×120 mm (Fig. 5.5).
Then:

Second moment of area of 'T' about NA = second moment of area of rectangle 100×130 about NA

Minus   2 × second moment of area of rectangle 45 × 120 about NA
or I_{NA} for 'T' = (I_{NA} for large rectangle + Ah^2) − 2 (I_{NA} for small rectangle + Ah^2)

$$= \left[\frac{bd^3}{12} + Ah^2 \right]_{\substack{large \\ rectangle}} - 2\left[\frac{bd^3}{12} + Ah^2 \right]_{\substack{small \\ rectangle}}$$

$$= \left[\frac{100 \times 130^3}{12} + (100 \times 130) \times 24 \cdot 545^2 \right]$$

$$-2\left[\frac{45 \times 120^3}{12} + 45 \times 120 \times 29 \cdot 545^2 \right]$$

$$= [18\ 308\ 333 + 7\ 831\ 941 \cdot 2] - 2[6\ 480\ 000 + 4\ 713\ 697 \cdot 9]$$
$$= 26\ 140\ 274 - 2 \times 11\ 193\ 697$$
$$= 3\ 752\ 888\ \text{mm}^4$$
$$\simeq \underline{3 \cdot 753 \times 10^{-6}\ \text{m}^4}$$

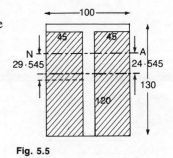

Fig. 5.5

The calculations are often best presented in tabular form.

Example 5.2

Calculate the position of the neutral axis and the second moment of area about that axis for the cross-section of the beam shown (Fig. 5.6).

Fig. 5.6

Shape	Area (mm²)	Distance of centroid from base XX (mm)	1st Moment of area about XX	2nd Moment of area about own NA $\left(\frac{bd^3}{12}\right)$	Distance of centroid from NA of whole (h)	h^2	Ah^2	$\frac{bd^3}{12}+Ah^2$
Rectangle 'A'	75 × 10 = 750	205	750 × 205 = 153750	$\frac{75 \times 10^3}{12}$ = 6250	111·7	12476·9	9357675	9363925
'B'	190 × 10 = 1900	105	1900 × 105 = 199500	$\frac{10\times190^3}{12}$ = 5715833	11·7	136·9	260110	5975943
'C'	120 × 10 = 1200	5	120 × 5 = 6000	$\frac{120\times10^3}{10}$ = 10000	88·3	7796·9	9356280	9366280
TOTAL	3850	$\bar{y}$	$3850\bar{y}$	I_{NA}	0	0	0	I_{NA}

To find $\bar{y}$:   153 750 + 199 500 + 6000 = $3850\bar{y}$
$$\therefore \bar{y} = 93 \cdot 3\ \text{mm}$$

Adding final column

$$I_{NA} = 9\,363\,925 + 5\,975\,943 + 9\,366\,290$$
$$= 24 \cdot 706148 \text{ mm}^4$$
$$\simeq \underline{24 \cdot 7 \times 10^{-6} \text{ m}^4}$$

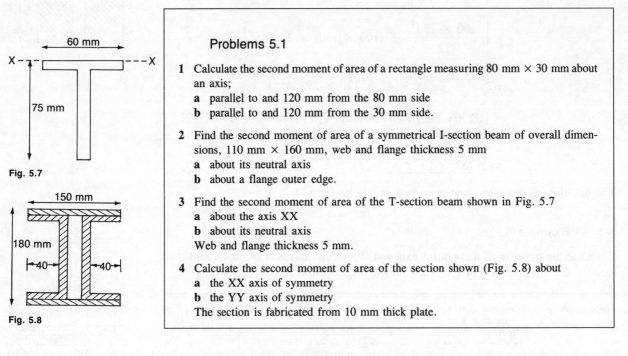

Fig. 5.7

Fig. 5.8

Problems 5.1

1 Calculate the second moment of area of a rectangle measuring 80 mm × 30 mm about an axis;
   a parallel to and 120 mm from the 80 mm side
   b parallel to and 120 mm from the 30 mm side.

2 Find the second moment of area of a symmetrical I-section beam of overall dimensions, 110 mm × 160 mm, web and flange thickness 5 mm
   a about its neutral axis
   b about a flange outer edge.

3 Find the second moment of area of the T-section beam shown in Fig. 5.7
   a about the axis XX
   b about its neutral axis
   Web and flange thickness 5 mm.

4 Calculate the second moment of area of the section shown (Fig. 5.8) about
   a the XX axis of symmetry
   b the YY axis of symmetry
   The section is fabricated from 10 mm thick plate.

5.2 Combined bending and direct stress

In previous study it has been shown that a bar or column may be subjected to a load which we assumed results in an even distribution of direct stress across its section. For example Fig. 5.9a shows a short column which is subjected to an applied compressive load F which gives rise to a compressive stress which, it is assumed, is evenly distributed across its cross-section, and which has a value of F/A.

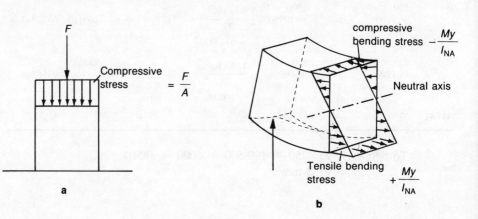

Fig. 5.9

It has also been shown that a beam subject to a bending moment, M, has a resulting stress pattern which consists of a compressive bending stress on the concave side of the beam, and a tensile bending stress on the convex side (Fig. 5.9b).

The numerical value of these bending stresses increases gradually with distance from the neutral axis and is a maximum at the outermost fibres of the beam. The value of the bending stress is given by the formula:

$$\sigma_b = \frac{M}{I}\, y$$

Where M is the applied bending moment, I the second moment of area of the section about the neutral axis, and y the distance measured from the neutral axis.

In nearly all cases we are concerned with the maximum bending stress, and this will occur at the outermost fibres, i.e. at y_{max}.

Then

$$\sigma_b = \frac{M}{Z} \quad \text{where} \quad Z = \frac{I}{y_{max}} = \text{the section modulus}$$

It is a common occurrence in engineering that a component or structure is subjected to both direct and bending stresses at the same time. The net result of this combination can be determined by considering the separate effects of the two types of stressing, and summing these effects algebraically.

For example, consider the short bar shown in Fig. 5.10. This bar is subjected to a direct tensile load F which will produce a tensile stress of F/A evenly distributed over

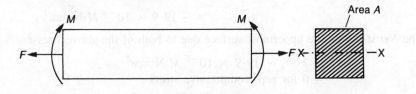

Fig. 5.10

the cross-sectional area A, together with a bending moment M which produces bending stresses of $-My/I$ (compressive) on the upper surface, and $+My/I$ (tensile) on the lower surface of the beam.

Adding these stresses algebraically results in a total stress value on the upper and lower surfaces as follows:

$$\text{Net stress on upper surface} = \frac{F}{A} - \frac{My}{I}$$

$$\text{Net stress on lower surface} = \frac{F}{A} + \frac{My}{I}$$

It is evident from the first equation that the net stress on the upper surface could be positive (tensile), zero, or negative (compressive) depending on whether F/A is greater, equal to, or less than My/I. It is also evident that the worst stress situation exists on the lower surface where the two component stresses are both tensile.

Example 5.3

A short bar of diameter 80 mm is subjected to an axial tensile load of 40 kN, calculate the greatest value of bending moment that may be applied so that there is no compressive stress in the bar.

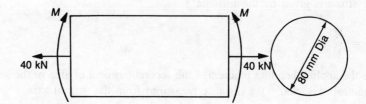

The direct tensile stress in bar $= \dfrac{F}{A}$

$$= \frac{40 \times 10^3}{\pi \, 80^2/4} = 7 \cdot 96 \text{ N/mm}^2$$

The bending stress on the uppermost surface of the bar due to the applied bending moment M

$$= -\frac{M \, d/2}{I} \text{ (compressive)}$$

$$= -\frac{M \, 80/2}{\pi \, 80^4/64}$$

$$= -19 \cdot 9 \times 10^{-6} \, M \text{ N/mm}^2$$

The *Net* stress on the uppermost surface due to both of the above stresses

$$= 7 \cdot 96 - 19 \cdot 9 \times 10^{-6} \, M \text{ N/mm}^2$$
$$= 0 \text{ for zero compressive stress}$$

$$\text{Thus } M = \frac{7 \cdot 96}{19 \cdot 9 \times 10^{-6}} = 4 \times 10^5 \text{ Nmm}$$

$$= \underline{400 \text{ Nm}}$$

Such a combination of bending and direct stresses occur in practice in several ways, two of which we will consider here.

Case one occurs when an axial load does not act through the centroid of the cross-sectional area of a beam, but is displaced from the true axis by some amount. In addition to the direct load there will be a bending moment equal to the value of the load multiplied by the amount that the load is displaced from the axis. The amount of displacement is called the **eccentricity**, and the load condition is called **eccentric loading** (Fig. 5.11).

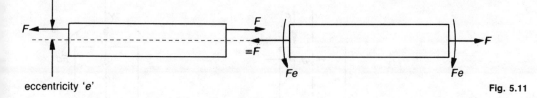

eccentricity 'e'   **Fig. 5.11**

A load F with an eccentricity e is equivalent to a load F with zero eccentricity together with a bending moment equal to $F \cdot e$.

The second case of combined bending and direct stresses occurs when axial and transverse loads result from an applied load being inclined to the longitudinal axis of a bar or beam (Fig. 5.12).

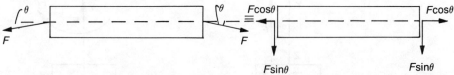

Fig. 5.12

A load F inclined at an angle θ to the true axial direction is equivalent to an axial load $F \cos \theta$, and a transverse load $F \sin\theta$. The transverse load causes a bending moment which, in this case, is a maximum at the centre of the beam. The two special cases of combined bending and direct stress we will now consider in greater detail.

5.3   Eccentric loading

Consider the short column shown in Fig. 5.13 subjected to an axial compressive load F which is displaced from the centroid of the cross-section by an amount d, measured in the x-direction in a three-dimensional co-ordinate system. Note that the load is only displaced from one axis (ZZ), and so only produces a bending moment about that axis.

The effect of the eccentric load F is to produce a direct compressive stress of value F/A evenly distributed over the cross-section, together with a bending stress My/Izz, due to the bending moment Fd, which will be tensile on the convex side, (the left-hand side in Fig. 5.13) and compressive on the concave side of the beam, (the right-hand side in Fig. 5.13). The resulting stress distribution will be a combination of the two separate stress distributions as illustrated in Fig. 5.14.

Examination of Fig. 5.14 shows that the stress distribution is predominantly compressive in this case, although a small tensile stress remains on the left-hand edge. Depending on the magnitudes of the overall compressive stress $-F/A$, and the bending stresses $+My/Izz$ and $-My/Izz$, this tensile stress could be zero as in Fig. 5.15 or it could, in fact, be a small compressive stress as in Fig. 5.16.

These last two cases have important engineering implications in that there are instances where a tensile force is extremely undesirable, for example in structures made from very brittle materials such as cast iron or concrete.

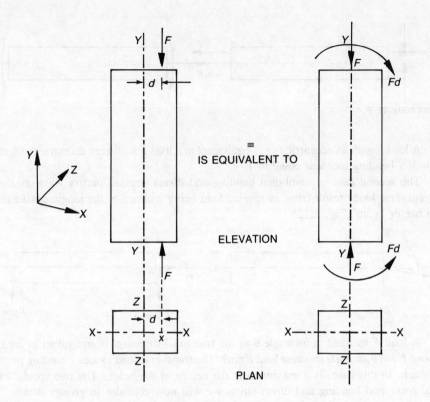

Fig. 5.13

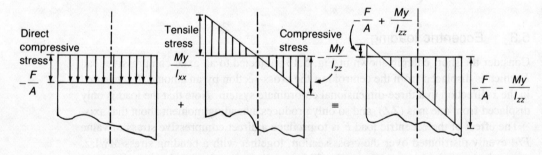

Fig. 5.14

Direct stress + Bending stress = Total stress

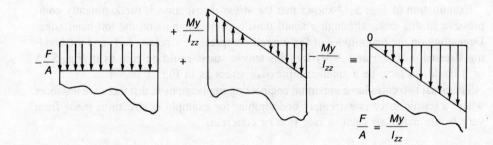

Fig. 5.15

Example 5.4

A concrete column of diameter 250 mm is acted upon by a longitudinal direct compressive load of 500 kN which is displaced 70 mm from the centroid of the cross-section. Calculate:

    a the magnitude of the maximum compressive and tensile stresses in the column;
    b the maximum eccentricity allowable for there to be no tensile stress in the column.

a The eccentric load of 500 kN is equivalent to an axial compressive load of 500 kN together with a bending moment of $500 \times 0\cdot070 = 35$ kNm.

The maximum compressive stress is given by

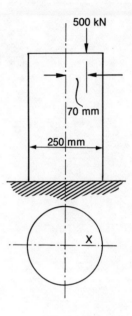

$$- F/A - \frac{My}{I}$$

$$= - \frac{500 \times 10^3}{\pi\, 0\cdot25^2/4} - \frac{35\cdot10^3 \times 0\cdot25 \times 64}{2\cdot\pi\,(0\cdot25)^+}$$

$$= - 10\cdot186 \times 10^6 - 22\cdot816 \times 10^6$$

$$= - 33 \text{ MN/m}^2$$

The maximum tensile stress is given by

$$- F/A + My/I$$

$$= - 10\cdot186 \times 10^6 + 22\cdot816 \times 10^6$$

$$= 12\cdot63 \text{ MN/m}^2$$

b For there to be no tensile stress in the column then $F/A \geq My/I$.

$$\text{or } \frac{F}{\pi\, d^2/4} \geq \frac{F.e.\ d/2}{\pi\, d^4/64}$$

which reduces to $e \leq d/8$

$$\text{thus } e \leq \frac{250}{8} \text{ mm}$$

$$e \leq 31\cdot25 \text{ mm}$$

Thus the maximum eccentricity allowable to avoid all tensile stress in the column is $31\cdot25$ mm.

Example 5.5

The steel G-clamp shown is designed to apply a clamping force up to 100 mm from the edge of the clamped material. At the section AB the clamp is in the form of an I-section beam, for which the second moment of area about the axis of bending is 5200 mm^4, and the cross-sectional area is 80 mm^2.

If the width of the cramp at AB is 25 mm, and the maximum allowable stress in the material at this section is 80 MN/m^2, calculate the maximum clamping force that can safely be applied.

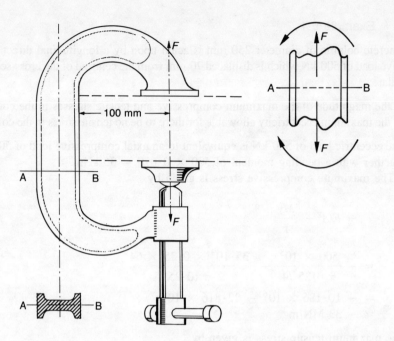

The total eccentricity of the load F is $112 \cdot 5$ mm from the axis of bending, and the effect of the load is to produce an overall tensile stress on the cross-section at AB, together with bending stresses due to a bending moment of value $F \times$ eccentricity (Fig. 5.16).

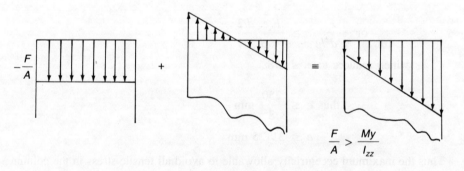

Fig. 5.16

The maximum stress will occur on the inside edge, B, of the clamp, where the overall tensile stress is added to by the tensile bending stress. Thus:

Maximum stress = direct tensile stress + tensile bending stress.

$$\text{or} \quad 80 \times 10^6 = \frac{F}{80 \times 10^{-6}} + \frac{F. \, 0 \cdot 1125 \, . \, 0 \cdot 0125}{5200 \times 10^{-12}}$$

$$80 \cdot 10^6 = F \, (12\,500 + 27\,043)$$

$$\text{when } F = 2 \cdot 02 \text{ kN}$$

Example 5.6

A short hollow column has a uniform rectangular section of external dimensions 75 mm × 100 mm and wall thickness 2·5 mm. The section carries a vertical compressive load of 30 kN acting at P which is offset from the neutral axis XX and YY by 15 mm and 20 mm respectively.

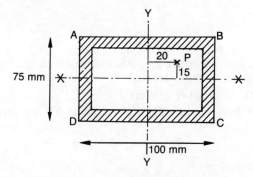

Calculate the stresses at the four corners A, B, C & D as a result of this loading.

Since the load has an eccentricity with respect to both axes of symmetry, then the column will experience resulting bending moments about both axes.

Bending moment about XX, Mxx = load × eccentricity
$$= 30 \times 10^3 \times 0\cdot015$$
$$= 450 \text{ Nm}$$

and, bending moment about YY, $Myy = 30 \times 10^3 \times 0\cdot020$
$$= 600 \text{ Nm.}$$

The second moment of area of the section about XX, Ixx is given by

$$Ixx = \frac{0\cdot1 \times 0\cdot075^3}{12} - \frac{0\cdot095 \times 0\cdot07^3}{12}$$
$$= 3\cdot516 \times 10^{-6} - 2\cdot715 \times 10^{-6}$$
$$= \underline{8 \times 10^{-7} \text{ m}}$$

and $$Iyy = \frac{0\cdot075 \times 0\cdot1^3}{12} - \frac{0\cdot07 \times 0\cdot095^3}{12}$$
$$= 6\cdot25 \times 10^{-6} - 5 \times 10^{-6}$$
$$= \underline{1\cdot25 \times 10^{-6} \text{ m}^4}$$

The whole cross-section will experience an overall compressive stress of value given by load divided by cross-sectional area.

Thus, direct compressive stress $= -\dfrac{30 \times 10^3}{(75 \times 100 - 70 \times 95) \times 10^{-6}}$
$$= -35\cdot3 \text{ MN/m}^2$$

The bending stresses that result from the bending moment about XX are given by

$$\pm \frac{Mxx\,y}{Ixx}$$

$$= \pm \frac{450.\ 0\cdot075/2}{8 \times 10^{-7}}$$

$$= \pm \underline{21\cdot1\ \text{MN/m}^2}$$

This bending stress will be compressive on the face AB and tensile on the face DC.
Thus, bending stress on face AB $= -21\cdot1\ \text{MN/m}^2$
and, bending stress on face CD $= +21\cdot1\ \text{MN/m}^2$
The bending stresses that result from the bending moment about YY are given by

$$\pm \frac{Myy.x}{Iyy}$$

$$= \pm \frac{600 \times 0\cdot1/2}{1\cdot25 \times 10^{-6}}$$

$$= \pm \underline{24\ \text{MN/m}^2}$$

This bending stress will be compressive on the face BC and tensile on the face AD.

Thus, bending stress on face BC $= -24\ \text{MN/m}^2$
and bending stress on face AD $= +24\ \text{MN/m}^2$
Now, each corner will experience the overall direct compressive stress plus the stresses
on its adjacent faces. Thus, net stress at corner A

$$= \text{direct stess} + \text{stress on face AB} + \text{stress on face AD}$$
$$= -35\cdot3 - 21\cdot1 + 24$$
$$= \underline{-32\cdot4\ \text{MN/m}^2}\ (\text{i.e. compressive})$$

the net stress at corner B

$$= \text{direct stress} + \text{stress on AB} + \text{stress on BC}$$
$$= -35\cdot3 - 21\cdot1 - 24$$
$$= \underline{-80\cdot4\ \text{MN/m}^2}\ (\text{i.e. compressive})$$

the net stress at corner C

$$= \text{direct stress} + \text{stress on BC} + \text{stress on CD}$$
$$= -35\cdot3 - 24 + 21\cdot1$$
$$= \underline{-38\cdot2\ \text{MN/m}^2}\ (\text{i.e. compressive})$$

the net stress at corner D

$$= \text{direct stress} + \text{stress on CD} + \text{stress on AD}$$
$$= -35\cdot3 + 21\cdot1 + 24$$
$$= \underline{+9\cdot8\ \text{MN/m}^2}\ (\text{i.e. tensile}).$$

Guided solution

A steel joist of 'I' section has the following dimensions: flange width 80 mm, overall depth 150 mm, web and flange thickness 10 mm. The joist is subjected to an axial tensile load of 45 kN applied at the centroid of the section. Over part of the joist's length it has been necessary to remove the lower flange. Calculate the increase in tensile stress that results from this action.

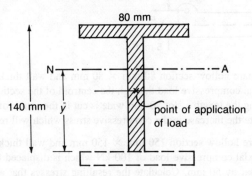

Step 1. Calculate the position of the neutral axis (NA) for the reduced section, by taking first moments of area about the lower end of the web ($\bar{y} = 91 \cdot 67$ mm)

Step 2. Calculate the value of I_{NA}, the second moment of area of the reduced section about NA, by using the standard formulae for rectangles and the parallel axis theorem if necessary ($I_{NA} = 2 \cdot 797 \times 10^{-6}$ m^4).

Step 3. Calculate the amount of eccentricity, i.e. the distance from the point of application of the load to the neutral axis of the reduced section (26·67 mm) and hence the value of the bending moment due to the eccentric load (1200·2 Nm).

Step 4. Calculate the maximum tensile bending stress which results from this bending moment, (39·3 MN/m^2) and the direct tensile stress that acts over all of the reduced section (21·43 MN/m^2).

Step 5. Calculate the direct tensile stress that acts over all of the full section (15·5 MN/m^2) and subtract this value from the sum of the answers to step 4 to give the required answer (45·23 MN/m).

Problems 5.2

1 A short steel strut of square section 50 mm × 50 mm carries an axial compressive load of 200 kN which is offset from the centroid of the section by 10 mm. Calculate the maximum compressive and tensile stresses in the strut, and sketch the stress distribution across the section.

2 A beam of 'I' section carries an axial tensile load of 25 kN at a point on the vertical axis of symmetry, but displaced from the centroid by an amount d. If the width of the flange is 75 mm, the overall depth 80 mm and the thickness of flange and web, 5 mm, and the maximum allowable tensile stress is 50 N/mm^2, calculate the maximum value of d.

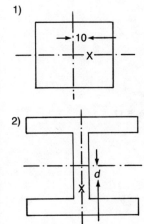

4)

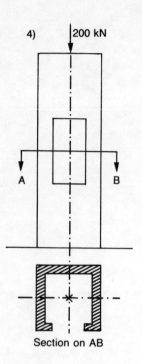

Section on AB

5)

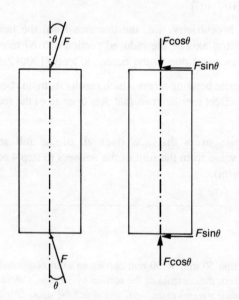

Fig. 5.17

3 A steel tie is fabricated from sheet 2·5 mm thick and is subjected to a tensile load applied at lugs giving an offset from the true axis of 6 mm. The tie is 10 mm wide and the maximum allowable stress is limited to 60 MN/m². Calculate the value of the maximum load which may be applied to the tie.

3)

4 A short steel column of square hollow section 80 mm × 80 mm and wall thickness 2 mm is subjected to an axial compressive load through the centroid of the section of 200 kN. At some point along the length a slot 50 mm wide is cut in the centre of one side of the column. Calculate the increase in the compressive stress which will result.

5 A cast-iron column of square hollow section 150 mm × 150 mm and wall thickness 25 mm is subjected to an axial compressive load of 100 kN which is displaced from the centroid along a diagonal by 50 mm. Calculate the resulting stresses that act at each of the outer four corners of the section.

5.4   Non-parallel loading

Consider the short column shown in Fig. 5.17 subjected to a compressive load F which is inclined at a small angle θ to the axial direction. This loading is equivalent to a true axial load of F cos θ, together with a transversely applied load F sin θ. The transverse load will give rise to a bending moment whose value will depend on the distance from the end of the column.

Example 5.7

A vertical steel tube $0 \cdot 3$ m long and internal and external diameters 25 mm and 30 mm respectively is built-in at its lower end, and is subjected at its upper end to a compressive load of 2 kN which is inclined at $5°$ to the vertical axis. Calculate the maximum compressive and tensile stresses in the column.

The maximum bending moment acts at the built-in end and has a numerical value of $2 \times 10^3 \sin 5° \times 0 \cdot 3 = 52 \cdot 29$ Nm. This bending moment causes a bending stress which is tensile on the right-hand edge of the column as viewed in the figure and compressive on the left-hand edge.

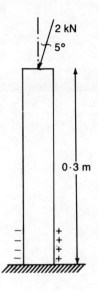

Thus, the bending stress $= \pm \dfrac{My}{I}$

$$= \pm \frac{52 \cdot 29 \times 0 \cdot 015}{\pi\,(0 \cdot 03^4 - 0 \cdot 025^4)/64}$$

$$= \frac{52 \cdot 29 \times 0 \cdot 015 \times 64}{\pi\,(8 \cdot 1 - 3 \cdot 9) \times 10^{-7}}$$

$$= \pm\ 38\ \text{MN/m}^2$$

The overall compressive stress $= \dfrac{\text{axial load}}{\text{cross-sectional area}}$

$$= -\frac{2000 \cos 5°}{\pi\,(0 \cdot 03^2 - 0 \cdot 025^2)/4}$$

$$= -\frac{2000.\ 0 \cdot 996\ .\ 4}{\pi\,(9 - 6 \cdot 25)10^{-4}}$$

$$= -\ 9 \cdot 22\ \text{MN/m}^2$$

The net stresses at the base of the column are thus

$$+\ 38 - 9 \cdot 22 = +28 \cdot 78\ \text{MN/m}^2 \ \text{(tensile on the right-hand edge)}$$
and $$-\ 38 - 9 \cdot 22 = -47 \cdot 22\ \text{MN/m}^2 \ \text{(compressive on the left-hand edge)}$$

Example 5.8

A concrete factory chimney may be regarded as a vertical column of hollow circular section, internal diameter 2 m and height 30 m. If the density of concrete is 1800 kg/m^3, and the structure must withstand a side force equivalent to a uniformly distributed load of 2000 N/m over its entire height, calculate the wall thickness necessary to avoid any tensile stress.

The compressive stress at the base of the chimney due to its own weight is given by

$$\frac{\text{load}}{\text{area}} = \frac{\text{density} \times \text{volume} \times \text{g}}{\text{area}}$$

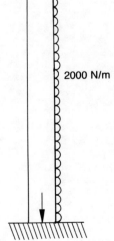

$$= \frac{\text{density} \times \text{area} \times \text{length} \times g}{\text{area}}$$

$$= \text{density} \times \text{length} \times g$$
$$= -1800 \times 30 \times 9 \cdot 81$$
$$= -0 \cdot 53 \text{ MN/m}^2$$

The uniformly distributed load is equivalent to a point load of $2000 \times 30 = 60\ 000$ N acting at the mid-height. Thus the bending moment at the base $= 60\ 000 \times 30/2 = 90\ 000$ Nm. This bending moment causes a bending stress which is tensile on the windward side and compressive on the other, and has a numerical value given by

$$\frac{My}{I} \quad \text{or} \quad \frac{M}{Z}$$

where Z is the section modulus.

Thus, bending stress $= \pm \dfrac{90\ 000}{\pi\ (D^3 - 2^3)/32}$

where D is the external diameter of the chimney. For there to be no tensile stress in the structure

$$\frac{90\ 000 \times 32}{\pi\ (D^3 - 2^3)} = 0 \cdot 53 \times 10^6$$

or $\qquad (D^3 - 8) = \dfrac{90\ 000 \times 32}{\pi. \ 0 \cdot 53 \times 10^6}$

$$= 1 \cdot 73$$

whence $D^3 = 9 \cdot 73$
and $\qquad D = 2 \cdot 135$ m

Thus the minimum wall thickness necessary is

$$\frac{0 \cdot 135}{2} = 0 \cdot 0675 \text{ m}$$

$$= \underline{67 \cdot 5 \text{ mm}}$$

Guided solution

The alloy mast of a racing yacht may be regarded as a vertical tube, 10 m long and of circular cross-section with inner and outer diameters 100 mm and 104 mm respectively. The mast is held in position by wire stays, the lateral ones of which are of negligible consideration. In the longitudinal plane however, there is a backstay which is attached to the masthead and angles at 40° to the mast, and a forestay which is attached at a point $2 \cdot 5$ m from the masthead and angled at 30° to the mast. Calculate the maximum tension that can be allowed in the backstay if the maximum allowable stress in the alloy mast is 40 MN/m^2.

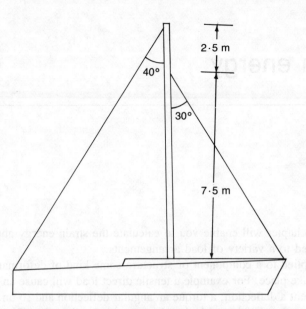

Step 1. Let the max. tension in the backstay be F, and calculate, in terms of F, what will then be the tension in the forestay. ($1 \cdot 414\ F$)

Step 2. By resolving the forces vertically, calculate the value of the overall compressive stress in the mast, again in terms of F. ($3 \cdot 015\ F\ \text{kN/m}^2$)

Step 3. By considering the horizontal components of the tensions calculate the net bending moment at the base of the mast. ($3 \cdot 321\ F\ \text{Nm}$)

Step 4. Calculate the maximum compressive bending stress at the base of the mast. ($4 \cdot 31\ F\ \text{N/m}^2$)

Step 5. Add the results of steps 2 and 4 and equate to the maximum allowable stress to find F. ($5 \cdot 46\ \text{kN}$)

Problems 5.3

1 A column is 200 mm high and of rectangular cross-section 100 mm × 50 mm. A 15 kN load is applied at the centre of the cross-section at an angle of 15° to the axis of the column. Calculate the value of the maximum tensile and compressive stresses in the column.

2 A vertical brick wall 10 m high is of uniform thickness for the whole of its height. The wall is designed to withstand a wind pressure of 1 kN/m², and the density of the brickwork is 2000 kg/m³. If all tensile stress is to be avoided, calculate the thickness of the wall.

3 A steel chimney is 50 m high 2 m external diameter and wall thickness 2 mm. If the density of the steel is 9000 kg/m³, and the chimney must withstand a horizontal wind pressure of 900 N/m², calculate the maximum stress in the steel.

1)

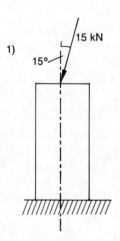

6 Strain energy

The contents of this chapter will enable you to calculate the strain energy absorbed by structures subjected to a variety of load arrangements.

When a load is applied to a component or structure some kind of deformation of the component will take place. For example a tensile direct load will cause an extension; a bending moment a deflection; a torque an angular deflection and so on. In all of these cases the point of application of the load moves with the distortion. We assume that the load is applied gradually so as to maintain an equilibrium situation at all times. In due course we will consider separately the case of a suddenly applied or impact load.

Now when a load or force moves through a distance work is done, where:

work done by load = load × distance moved by load.

The work that is done is absorbed by the component as stored or **strain energy**. Providing that the component has not been deformed beyond its elastic limit it will use this strain energy to restore itself to its original state once the load is removed. In other words the component behaves rather like a spring. Thus:

work done by load = strain energy absorbed

6.1 Strain energy due to direct loading

Suppose that an increasing load F is gradually applied to a bar of cross-sectional area A, and length ℓ. The load will produce an extension which is always proportional to the load (Fig. 6.1b). The work done by the load is equal to the area under the load-extension graph, i.e.

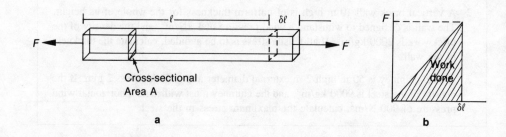

Fig. 6.1

a

b

$$\text{work done} = \tfrac{1}{2}\,.\,F.\ \delta\ell \tag{6.1}$$

Also
$$E = \frac{\text{stress}}{\text{strain}} = \frac{F/A}{\delta\ell/\ell}$$

re-arranging
$$\delta\ell = \frac{F\ell}{EA} \tag{6.2}$$

substituting (6.2) into (6.1).

$$\text{work done} = \tfrac{1}{2}\,.\,F\,.\,\frac{F\ell}{EA}$$

$$= \frac{F^2\ell}{2\ EA}$$

but the strain energy, U = work done

$$\therefore U = \frac{F^2\ell}{2\ EA} \tag{6.3}$$

The strain energy expression can also be expressed in terms of the applied stress since

$$\sigma = F/A$$

or
$$F^2 = \sigma^2\,A^2$$

Thus (6.3) becomes

$$U = \frac{\sigma^2}{2E}\,A\ell$$

The product $A\ell$ is the volume of the component so:

$$U = \frac{\sigma^2}{2E} \text{ per unit volume} \tag{6.4}$$

Note that since σ is squared this expression is the same for both the tensile and compressive cases.

Example 6.1

Calculate the strain energy absorbed by a steel bar of diameter 50 mm, length $0\cdot75$ m when subjected to a tensile load of 100 kN. If $0\cdot25$ m of the bar is subsequently turned down to 25 mm calculate the new strain energy absorbed by the bar. $E = 200$ GN/m^2.

a From equation (6.3)

$$U = \frac{F^2\ell}{2\ EA}$$

$$= \frac{(100 \times 10^3)^2 \times 0\cdot75 \times 4}{2 \times 200 \times 10^9 \times \pi \, (0\cdot05)^2}$$

$$= \underline{9\cdot55 \text{ joules}}$$

b The strain energy absorbed by $0\cdot5$ m of bar, $U_{0\cdot5}$ is given by:

$$U_{0\cdot5} = \frac{F^2\ell}{2 \, EA} = \frac{(100 \times 10^3)^2 \times 0\cdot5 \times 4}{2 \times 200 \times 10^9 \times \pi \, (0\cdot05)^2}$$

$$= \underline{6\cdot366 \text{ joules}}$$

and the strain energy absorbed by the remaining $0\cdot25$ m, $U_{0\cdot25}$ is given by

$$U_{0\cdot25} = \frac{F^2\ell}{2 \, EA} = \frac{(100 \times 10^3)^2 \times 0\cdot25 \times 4}{2 \times 200 \times 10^9 \times \pi \, (0\cdot025)^2}$$

$$= \underline{12\cdot73 \text{ joules}}$$

The total strain energy absorbed, $U = U_{0\cdot5} + U_{0\cdot25}$

$$= 6\cdot366 + 12\cdot73$$

$$= \underline{19\cdot1 \text{ joules}}$$

The strain energy absorbed by the turned down bar is twice that of the original.

Problems 6.1

1   A steel bar 250 mm long of square section of side 25 mm is subjected to a compressive load of 20 kN, calculate the amount of strain energy absorbed. $E = 200 \text{ GN/m}^2$.

2   If the bar in question 1 is drilled out with a hole 10 mm diameter, co-axial with the bar for 100 mm of its length, calculate the new strain energy absorbed by the bar.

3   The maximum stress allowable in a steel bar is 160 MN/m². The bar is $0\cdot74$ m long and has a diameter of 75 mm over the first $0\cdot5$ m and a diameter of 50 mm over the remainder of its length. If $E = 208 \text{ GN/m}$ calculate the strain energy capacity of the bar.

4   Calculate the direct stress caused in a mild steel bar $0\cdot4$ m long and 15 mm diameter if 10 joules of work done in stretching it.

5   Failure of a material occurs when the total strain energy absorbed reaches 200 kJ/m³. A component made of the material is subjected to principal stresses σ_1 and σ_2. If both principal stresses are tensile and $\sigma_1 = 180 \text{ MN/m}^2$ calculate the maximum value of σ_2. $E = 200 \text{ GN/m}^2$ and $\nu = 0\cdot3$.

6.2   Strain energy due to shear

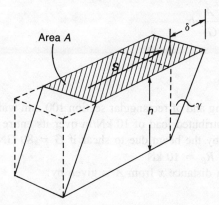

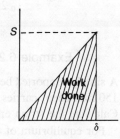

Fig. 6.2

A shear force S will produce a deflection δ (Fig. 6.2) where:

$$\gamma = \frac{\delta}{h} \text{ radians}$$

and γ is the shear strain.

If S is gradually applied, and its point of application moves a distance δ then

$$\text{strain energy due to shear} = \tfrac{1}{2} . S . \delta$$

since

$$G = \frac{\text{shear stress}}{\text{shear strain}} = \frac{\tau}{\gamma} = \frac{Sh}{A\delta}$$

then the strain energy due to shear $= \tfrac{1}{2} . S . \dfrac{Sh}{AG}$

$$= \frac{S^2 h}{2\,AG} \tag{6.5}$$

The shear stress $\tau = S/A$, hence $S^2 = \tau^2 A^2$ whence

$$U = \tfrac{1}{2} . \frac{\tau^2 A^2 h}{GA}$$

$$= \frac{\tau^2}{2G} . Ah$$

$$= \frac{\tau^2}{2G} \times \text{volume of solid} \tag{6.6}$$

The similarity between equations (6.6) and (6.4) is readily apparent.

In general the strain energy absorbed by a bar or beam due to shear forces will vary at each section along the beam. For a beam of length ℓ the general expression for the total strain energy absorbed by the whole beam due to shear will be:

$$U_{total} = \int_0^\ell \frac{S^2}{2GA} \, dx = \int_0^\ell \frac{\tau^2 A}{2G} \, dx \tag{6.7}$$

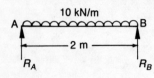

10 kN/m

A ⎯⎯⎯⎯⎯⎯⎯ B

—— 2 m ——

R_A R_B

Example 6.2

A simply supported beam AB, 2 m long and of rectangular section 100 mm wide and 150 mm deep, carries a uniformly distributed load of 10 kN/m over its entire span. Calculate the strain energy absorbed by the beam due to shear if $G = 80 \text{ GN/m}^2$.

For equilibrium of the beam $R_A = R_B = 10$ kN

The shear force at any section xx a distance x from A is given by:

$$S = 10x - 10$$
$$= 10 \, (x-1) \text{ kN}$$

The strain energy absorbed by the beam,

$$U = \int_0^2 \frac{[10(x-1)]^2}{2 \, GA} \, dx$$

$$= \frac{100}{2 \, GA} \int_0^2 (x-1)^2 \, dx$$

$$= \frac{100}{2 \, GA} \left[\frac{(x-1)^3}{3} \right]_0^2$$

$$= \frac{100}{2 \, GA} \cdot 2/3$$

Thus $$U = \frac{100 \times 2}{2 \times 80 \times 10^9 \times 0 \cdot 1 \times 0 \cdot 15 \times 3}$$

$$= \underline{2 \cdot 78 \times 10^{-8} \text{ kJ}}$$

In a practical situation such as the beam just described it is found that the strain energy due to shear is usually negligible compared with the strain energy due to bending.

Problems 6.2

1 For the beams shown in Problems 2.7 numbers **1a, b, c** and **d** calculate the strain energy absorbed due to shear only. Assume in all cases that $G = 70 \text{ GN/m}^2$, and that the area of cross-section is $0 \cdot 02 \text{ m}^2$.

6.3   Strain energy due to bending

A bending moment will produce a deformation in the form of a change of slope or radius of curvature. For example consider a small element of a beam of length δx which is subjected to a gradually applied bending moment M.

The work done by the bending moment is equal to the strain energy absorbed by the element, or:

$$\delta U = \tfrac{1}{2} . M . \delta\theta$$

Since $\delta x = R\delta\theta$ where R is the radius of curvature then

$$\delta U = \tfrac{1}{2} M \frac{\delta x}{R}$$

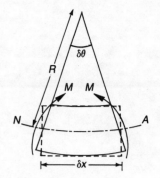

Fig. 6.3

From the bending beam formula $\dfrac{1}{R} = \dfrac{M}{EI}$

$$\therefore \delta U = \frac{M^2}{2\,EI}\,\delta x$$

For a beam of length ℓ, the total strain energy is found by integrating thus:

$$U = \int_0^\ell \frac{M^2}{2EI}\,dx \tag{6.8}$$

Example 6.3

Find the strain energy absorbed by the beam in Example 6.2 due to the bending moment alone. $E = 200$ GN/m^2.

The bending moment expression at any section xx a distance x from A is:

$$M = 10x - 10x . x/2$$
$$= 10x - 5x^2$$

The strain energy due to bending is given by:

$$U = \int_0^2 \frac{M^2}{2EI}\,dx$$

$$= \int_0^2 \frac{(10x - 5x^2)^2}{2EI}\,dx$$

$$= \int_0^2 \frac{(100x^2 - 100x^3 + 25x^4)}{2EI}\,dx$$

$$= \frac{1}{2EI}\left[\frac{(100x^3)}{3} - 25x^4 + 5x^5 \right]_0^2$$

$$= \frac{1}{2EI}\left(\frac{800}{3} - 400 + 160\right)$$

$$= \frac{13\cdot 3}{EI}$$

$$= \frac{13\cdot 3 \times 12}{200 \times 10^9 \times 0\cdot 1 \times (0\cdot 15)^3}$$

$$= \underline{2\cdot 37 \times 10^{-6}\ \text{kJ}}$$

Problems 6.3

1 For the same beams as in Problems 2.7, **1a, b, c** and **d** calculate the strain energy absorbed due to bending moment only. Assume in all cases that $E = 200\ \text{GN/m}^2$ and $I = 6\cdot 67 \times 10^{-5}\ \text{m}^4$.

6.4   Strain energy due to torsion

A gradually applied torque will produce a deformation in the form of a twist. For example consider an elemental length of a shaft, δx which suffers a twist of $\delta\theta$ due to a gradually applied torque T, (Fig. 6.4).

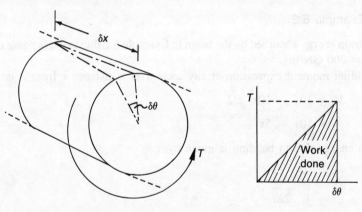

Fig. 6.4

The work done in twisting = strain energy stored

or $\frac{1}{2}\cdot T\ \delta\theta = \delta U$

From the torsion of shafts formula

$$\frac{T}{J} = \frac{G\delta\theta}{\delta x}$$

or $\delta\theta = \frac{T\delta x}{GJ}$

Thus $\quad \delta U = \dfrac{T^2 \delta x}{2\ GJ}$

For a shaft of length ℓ, the total strain energy U is thus

$$U = \int_0^\ell \dfrac{T^2 dx}{2\ GJ} \tag{6.9}$$

For a torque which is constant along a shaft this expression reduces to

$$U = \dfrac{T^2 \ell}{2\ GJ}$$

for a shaft of diameter d, $J = \dfrac{\pi d^4}{32}$

and $\quad T = \dfrac{\tau J}{d/2}$

Then

$$U \pm \left(\dfrac{\tau J}{d/2} \right)^2 \dfrac{\ell}{2\,GJ}$$

$$= \dfrac{2\tau^2 J \ell}{d^2 G}$$

$$= \dfrac{2\tau^2 \pi d^4 \ell}{d^2\ 32G}$$

$$= \dfrac{\tau^2}{4G} \times \dfrac{\pi d^2 \ell}{4}$$

$$= \dfrac{\tau^2}{4G} \times \text{volume of bar} \tag{6.10}$$

Example 6.4

A hollow shaft is $0 \cdot 5$ m long and has an external diameter of 75 mm and an internal diameter of 65 mm. The shaft is gradually subjected to a torque of 10 kNm and the modulus of rigidity of the material shaft is 80 GN/m^2. Using strain energy considerations calculate the degree of twist suffered by the shaft due to the torque. Check your answer by using the simple torsion of shafts formula.

a The work done by the torque = Strain energy absorbed by bar

or $\quad \frac{1}{2} \cdot T \cdot \theta = \displaystyle\int_0^{0 \cdot 5} \dfrac{T^2}{2\ GJ}\ dx$

$$\tfrac{1}{2} \cdot 10 \times 10^3 \cdot \theta = \frac{(10 \cdot 10^3)^2 \; [x]_0^{0 \cdot 5}}{2 \; GJ}$$

$$\theta = \frac{10 \times 10^3 \times 0 \cdot 5 \times 32}{80 \times 10^9 \times \pi \; (0 \cdot 075^4 - 0 \cdot 065^4)}$$

$$= \underline{0 \cdot 0462 \text{ rads}}$$

b Check using torsion formula $\dfrac{T}{J} = \dfrac{G\theta}{\ell}$

or $\theta = \dfrac{T\ell}{GJ}$

$$= \frac{10 \times 10^3 \times 0 \cdot 5 \times 32}{80 \times 10^9 \times \pi \; (0 \cdot 075^4 - 0 \cdot 065^4)}$$

$$= \underline{0 \cdot 0462 \text{ rads}}$$

Example 6.5

A cranked bar ABC of 25 mm diameter lies in a horizontal plane with AB = 0·6 m, BC = 0·4 m and $\angle$ABC = 90°. The bar is rigidly built in at A, and a load of 800 N is carried at the free end C. If $E = 208$ GN/m^2, and $G = 70$ GN/m^2, calculate the vertical deflection at the free end.

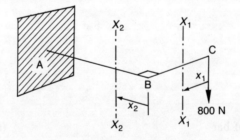

The portion of the bar AB is subjected to shear, bending and torsion where:

shear force, $S = 800$ N.
bending moment, $M = 800x_2$ where x_2 is the distance along AB from B
and torque, $T = 800 \times 0 \cdot 4 = 320$ Nm.

The strain energy absorbed by AB is the sum of the strain energies due to shear, bending and torsion

i.e. $$U_{AB} = \int_0^{0 \cdot 6} \frac{S^2}{2 \, GA} \, dx_2 + \int_0^{0 \cdot 6} \frac{M^2}{2 \, EI} \, dx_2 + \int_0^{0 \cdot 6} \frac{T^2}{2 \, GJ} \, dx_2$$

$$= \frac{S^2}{2 \, GA} \, [x_2]_0^{0 \cdot 6} + \frac{800^2}{2 \, EI} \left[\frac{x_2^3}{3} \right]_0^{0 \cdot 6} + \frac{T^2}{2 \, GJ} \, [x]_0^{0 \cdot 6}$$

$$= \frac{800^2 \times 0.6 \times 4}{2 \times 70 \times 10^9 \times \pi \times 0.025^2}$$

$$+ \frac{800^2 \times 0.6^3 \times 64}{2 \times 208 \times 10^9 \times \pi \times 0.025^4 \times 3}$$

$$+ \frac{320^2 \times 0.6 \times 32}{2 \times 70 \times 10^9 \times \pi \times (0.025)^4}$$

$$= 5.588 \times 10^{-3} + 5.777 + 11.444$$

$$= \underline{17.23 \text{ joules}}$$

The portion of the bar BC is subjected to shear and bending. As demonstrated for the portion AB the strain energy due to shear is negligibly small.

Thus $U_{BC} = \displaystyle\int_0^{0.4} \frac{M^2}{2EI} \, dx_1$

The bending moment, M at a section xx_1 a distance x_1 from C is given by: $M = 800 \, x_1$

Thus $U_{BC} = \displaystyle\int_0^{0.4} \frac{(800 \, x_1)^2 \, dx}{2 \, EI} = \frac{800^2}{2EI} \left[\frac{x_1^3}{3} \right]_0^{0.4}$

$$= \frac{800^2 \times 0.4^3 \times 64}{2 \times 208 \times 10^9 \times \pi \times 0.025^4 \times 3}$$

$$= \underline{1.712 \text{ joules}}$$

The total strain energy absorbed by the bar is

$$U = U_{AB} + U_{BC}$$

$$= 17.23 + 1.712$$

$$= \underline{18.94 \text{ joules}}$$

The work done by the load = total strain energy absorbed

or $\frac{1}{2} . 800 . \delta = 18.94$

$$\therefore \delta = \underline{47.35 \text{ mm}}$$

Problems 6.4

1   A stepped shaft has a total length of 0.5 m. The diameter of the first 0.3 m is 50 mm, and the diameter of the remainder of its length is 40 mm. A torque of 650 Nm is applied to one end of the shaft the other end being held rigidly. Calculate the strain energy absorbed by the shaft, and the angle that one end turns through relative to the other. Take $G = 72$ GN/m^2.

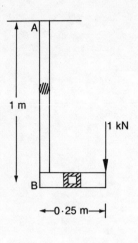

1 m

B

←— 0·25 m —→

A

1 kN

2 A cranked bar ABC is fabricated from a hollow round section AB 1 m long with external diameter 30 mm and internal diameter 25 mm welded at right-angles to a hollow square section bar BC of side 30 mm and wall thickness 25 mm, and length 0·25 m. The cranked bar is mounted in a vertical plane and rigidly built-in at A. Calculate the vertical deflection of the free end C if at this point a 1 kN load is applied. Neglect the strain energy due to shear and tension and take $E = 200$ GN/m^2.

3 If the bar in question **2** is mounted horizontally and the load of 1 kN still applied vertically at C, calculate the vertical deflection of C. Neglect the strain energy due to shear and take $G = 80$ GN/m^2.

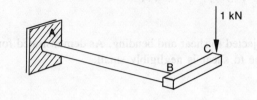

1 kN

A

C

B

6.5   Impact loading

The strain energy examples that have been solved up to this point have been careful to state that the loads are applied gradually. A gradually applied load ensures that a state of equilibrium is maintained at all times, and that the energy required to be absorbed by the material is only that due to the load moving its point of application.

Invariably if the load is applied suddenly as in an impact situation the initial deflection of the member is more than it would have been had the same load been applied gradually. There is also usually some oscillation of the load about its final resting position associated with an impact. The extra deflection is due to the kinetic energy possessed by a load that is moving. The strain energy absorbed by the material is greatest at the maximum displacement, at which point all the energy of the load has been given up. It is assumed that there is no loss of energy at impact.

6.6   Impact loading inducing tensile stress

Consider a load F which is allowed to fall a distance h onto a horizontal platform which is rigidly attached to a rod of length ℓ and area A, built in at its uppermost end (Fig. 6.5). The weight of the rod and platform is assumed to be negligible in comparison with the load F so that it can be assumed that there is no loss of energy at impact.

As the load impacts with the platform the rod will experience a tensile stress and an extension.

Let the maximum extension be x, so that the total distance moved by the load will be $(h+x)$.

The potential energy lost by the load = the gain in strain energy of the rod.

or
$$F(h+x) = \frac{\sigma^2}{2E} \times A\ell \qquad (6.11)$$

Fig. 6.5

ℓ

h

x

replacing x by $\dfrac{\sigma\ell}{E}$ will give a quadratic equation in σ whose positive root will be the required tensile stress.

If the load F is held *in contact* with the platform and released, the case of a suddenly applied load without impact, then

$$h = 0 \text{ in equation 6.11.}$$

Thus $Fx = \dfrac{\sigma^2}{2\,E} \times A\ell$

replacing x with $\dfrac{\sigma\ell}{E}$ gives

$$\frac{F\sigma\ell}{E} = \frac{\sigma^2}{2\,E} \times A\ell$$

whence $\sigma = 2\,F/A$ (6.12)

i.e. the maximum stress induced by a suddenly applied load is *twice* that produced by the same load applied gradually.

Example 6.6

A collar of weight 150 N falls 20 mm onto a platform which is rigidly attached to an alloy rod of length 250 mm and cross-sectional area 4 mm². If E for the alloy is 100 GN/m² calculate the maximum stress induced in the rod and the maximum deflection.

Let σ_{max} N/m² be the maximum stress

The maximum load in the rod $= \sigma_{max} \times$ area

$$= 4 \times 10^{-6}\, \sigma_{max} \text{ N}$$

and the maximum extension, $x = \dfrac{\sigma_{max}}{E} \times \ell$

$$= \frac{\sigma_{max} \times 0\cdot25}{100 \times 10^9} \text{ m.}$$

The potential energy lost = strain energy gained or

$$150 \left(0\cdot02 + \frac{\sigma_{max} \times 0\cdot25}{100 \times 10^9} \right) = \tfrac{1}{2}(4 \times 10^{-6}\, \sigma_{max}) \left(\frac{\sigma_{max} \times 0\cdot25}{100 \times 10^9} \right)$$

$$3 \times 10^{11} + 37\cdot5\, \sigma_{max} = 0\cdot5 \times 10^{-6}\, \sigma_{max}^2$$

$$\sigma_{max}^2 - 75 \times 10^6\, \sigma_{max} - 6 \times 10^{17} = 0$$

whence $\sigma_{max} = 813$ MN/m² or -737 MN/m².

The negative result gives the compressive stress on rebound assuming the load stays in contact with the platform.

The maximum tensile stress is then 813 MN/m²

The maximum deflection, $x = \dfrac{\sigma_{max} \times 0.25}{10^{11}}$ m

$$= \frac{813 \times 10^6 \times 0.25 \times 10^3 \text{ mm}}{10^{11}}$$

$$= \underline{2.03 \text{ mm}}$$

6.7 Impact loading inducing torsion shear stresses

A similar impact situation arises in the case of shafts rotating at speed which are suddenly stopped perhaps by seizure of bearings or gear failure. The sudden loss in kinetic energy equals the gain in strain energy of shaft.

Example 6.7

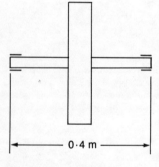

0.4 m

A drive shaft consists of a solid bar of diameter 75 mm supported in bearings 0.4 m apart. The shaft carries a flywheel at its centre span of mass 20 kg and radius of gyration 0.12 m. The flywheel is rotating at 350 rev/min when a gear failure causes the shaft to suddenly jam. Calculate the increase in stress induced in the shaft by the failure. Neglect the inertia of the shaft, and take $G = 70$ GN/m².

The stress due to the sudden additional torque is all that is required.

Moment of inertia of flywheel, $I = mk^2$
$$= 20 \, (0.12)^2$$
$$= \underline{0.288 \text{ kg m}^2}$$

The kinetic energy of the flywheel $= \frac{1}{2} . I \, \omega^2$

$$= \frac{1}{2} \times 0.288 \times \left(\frac{350 \, . \, 2\pi}{60}\right)^2$$

$$= \underline{193.4 \text{ joules}}$$

The loss in KE = gain in strain energy of shaft

$$= \frac{\tau^2}{4 \, G} \times \frac{\pi d^2 \ell}{4} \text{ from equation (6.10)}$$

Thus $193.4 = \dfrac{\tau^2 \times \pi \, (0.075)^2 \times 0.4}{4 \times 70 \times 10^9 \times 4}$

$$= \frac{\tau^2}{1.58 \times 10^{14}}$$

whence $\tau = 174.8$ MN/m²

Thus the shear stress induced by the sudden stopping of the shaft is 174.8 MN/m².

Guided solution

A cantilever of length 2 m and circular cross-section 50 mm diameter is subjected to a gradually applied load of 600 N at its free end. Calculate the vertical deflection at the free end. If the load is dropped onto the free end from a height of 15 mm what will then be the maximum deflection and the value of the maximum bending stress in the cantilever? $E = 200$ GN/m^2.

a

Step 1. Write down an expression for the bending moment at a section a distance x from the free end, and substitute this into the standard formula for the strain energy due to bending.

Step 2. Equate this expression to the work done by the load in moving its point of application through the deflection d and evaluate the deflection.   (26 mm)

b

Step 1. Let the maximum stress in the cantilever be σ_{max} and hence deduce an expression for the effective load We at the free end that would cause this stress using the relationship

$$\frac{M}{I} = \frac{\sigma_{max}}{d/2} \text{ where } M = W_e \times 2 \qquad (W_e = 6\cdot136 \times 10^{-6} \sigma_{max})$$

Step 2. Let the maximum deflection of the free end of the cantilever be d metres, and deduce an expression for d in terms of σ_{max} using the known relationship

$$d = \frac{W_e \ell^3}{3\,EI} \qquad \text{(see page 189)}$$

n.b. d is equal to the deflection that would be caused by W_e applied gradually
$$(d = 2\cdot67 \times 10^{-10} \sigma_{max})$$

Step 3. The potential energy lost by the 600 N load in falling through a total distance of $(0\cdot015 + d)$ metres is equal to the strain energy stored, which in turn is equivalent to the work done by the effective load W_e in moving through a distance d. Set up the equation and solve the resulting quadratic equation in σ_{max} taking the positive root. $(\sigma_{max} = 241$ MN/m$^2)$

Step 4. Substituting for σ_{max} in the expression for maximum deflection to give the result required. $(d = 64$ mm)

Problems 6.5

1 A vertical steel bar is rigidly fixed at its upper end, and carries a platform rigidly fixed to its lower end. The bar is $0\cdot5$ m long and has a diameter of 25 mm. If a weight of 500 N is dropped from a height of 20 mm onto the platform calculate the maximum stress induced in the bar, and the maximum extension. $(E = 200$ GN/m$^2)$

2 **a** For the bar in question **1** determine from what height a weight of 500 N may be dropped onto the platform if the maximum allowable stress in the bar is 100 MN/m^2.

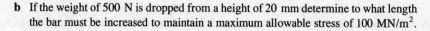

b If the weight of 500 N is dropped from a height of 20 mm determine to what length the bar must be increased to maintain a maximum allowable stress of 100 MN/m^2.

3 A circular steel shaft of length 1·6 m and diameter 50 mm is fitted with a flywheel, of mass 50 kg and radius of gyration 0·25 m, at one end. The flywheel is rotating at 120 rev/min when the other end of the shaft is suddenly stopped. Determine the maximum stress that occurs in the shaft due to the impact. $G = 70$ Gn/m^2

4 A winding drum is driven by a solid shaft 0·25 m long, and 25 mm diameter. The drum is 0·6 m diameter, has a mass of 20 kg and a radius of gyration of 0·2 m. The drum is used to raise a load at a speed of 1 m/s via a cable which is wound onto the drum. At a point in the lifting operation the load is caught and suddenly stopped. Assuming that the cable does not break, and that there is negligible stretch, determine the maximum shear stress induced in the shaft. ($G = 80$ GN/m^2)

5 A simply supported beam 2 m long is of rectangular section 150 mm wide by 200 mm deep. A load of 400 N falling from a height of 50 mm strikes the beam at the centre of its span. Calculate the maximum deflection of the beam. Take E for the material of the beam to be 15 GN/m^2.

6 Calculate the section modulus of an 'I' section centilever beam of depth 100 mm required to withstand a load of 250 N falling from a height of 75 mm onto its free end. The length of the cantilever is 3 m and the maximum allowable stress in the beam is 200 MN/m^2.   ($E = 200$ GN/m^2)

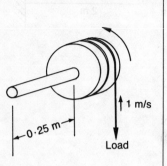

1 m/s

—0·25 m—

Load

7 Beam analysis

The contents of this chapter will enable the student to calculate the slope and deflection of any part of cantilever and simply supported beams carrying point or U.D. loads, and to derive the relationships between intensity of loading, shearing force and bending moment.

7.1 The relationship between loading, shear force and bending moment

A complete analysis of shear force and bending moment and the sketching of the associated diagrams has been covered in Chapter 2. The nature of shear force and bending moment and the calculation of expressions representing them at different portions of the beam leads to an analysis of the relationships that exist between these expressions. Such analysis can lead an experienced engineer to obtain shear force and bending moment diagrams, and particularly peak values, points of contraflexure and so on.

Consider an elemental length of beam, δx distance x from the left-hand end of the beam and carrying a uniformly distributed load of w per unit length (Fig. 7.1). At distance x the shear force is F and the bending moment M. At distance $x+\delta x$ the shear force is $F+\delta F$ and the bending moment $M+\delta M$.

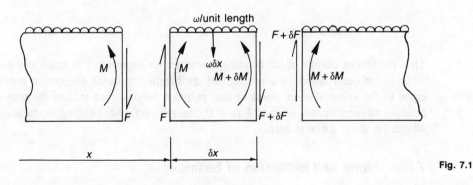

Fig. 7.1

For equilibrium of the element, resolving vertically

$$F + \delta F = F - w\delta x$$

$$\text{or} \quad \frac{\delta F}{\delta x} = -w$$

In the limit, as $\delta x \to 0$

$$\frac{dF}{dx} = -w$$

Integrating with respect to x

$$F = -\int w dx$$

In words, the rate of change of shear force with respect to x is equal to the load intensity. With reference to the shear force diagram, the gradient of the diagram at any point is equal to the load intensity at that point. Also the value of shear force at a point is equal to the integral of the load intensity at that point.

Also, for equilibrium of moments for the element taking moments about the left hand edge of the element,

$$M + \delta M = M + w.\delta x. \frac{\delta x}{2} + (F + \delta F)\delta x$$

Ignoring the product of small quantities

$$\delta M = F \delta x$$

or $$\frac{\delta M}{\delta x} = F$$

In the limit as $\delta x \to 0$

$$\frac{dM}{dx} = F$$

Integrating with respect to x

$$M = \int F dx$$

Thus the rate of change of the bending moment with respect to x is equal to the shear force or, in other words, the gradient of the bending moment diagram at a point is equal to the value of shear force at that point. It follows then that as the maximum bending moment occurs when $dM/dx = 0$ then the maximum bending moment occurs where the shear force is zero.

7.2   Slope and deflection of beams

In engineering structures subjected to external loads there will always be some deflection of the members of the structure. In many cases the deflection will be small and of no particular consequence, but in other cases, for example in aircraft structures and precision machines, the size of the deflections are an important and necessary piece of information to the engineer. The theory that follows allows the deflections of simple beams to be calculated.

In Chapter 2 it was shown how the bending moment at any section a distance x from the left-hand end of a loaded beam can be calculated. If the extreme left-hand end of the beam is retained as the origin of co-ordinates in a rectangular system, then, as before x will be the distance measured along the beam, and y will represent the size of the deflection (Fig. 7.2).

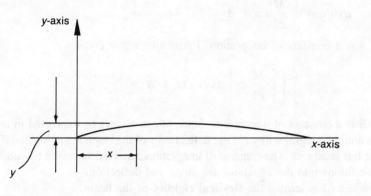

Fig. 7.2

Note that y will be positive along portions of the beam deflected above the unloaded, undeflected position, usually a hogging beam, and negative along portions of the beam deflected below the neutral position, usually a sagging beam.

The shape of a beam described in such an $x-y$ co-ordinate system could be represented by an equation of the form $y = f(x)$. Differentiation of the function y will give the slope of the beam as a function of x, and further differentiation will give d^2y/dx^2 the second differential coefficient of y with respect to x.

Now it can be shown, mathematically, that

$$R = \frac{[1 + (dy/dx)^2]^{3/2}}{d^2y/dx^2}$$

where R is the radius of curvature of the beam. Since dy/dx is nearly always very small then $(dy/dx)^2$ can be ignored and so the relationship reduces to

$$R \simeq \frac{1}{d^2y/dx^2}$$

or
$$\frac{1}{R} = d^2y/dx^2 \tag{7.1}$$

From simple bending theory

$$\frac{M}{I} = \frac{E}{R}$$

or
$$\frac{1}{R} = \frac{M}{EI} \tag{7.2}$$

Equating (7.1) and (7.2) gives

$$d^2y/dx^2 = \frac{M}{EI}$$ (7.3)

Integrating equation (7.3) with respect to x gives

$$dy/dx = \int \frac{M}{EI} \, dx + A$$

where A is a constant of integration. Integrating again gives

$$y = \int \left[\int \frac{M}{EI} \, dx \right] dx + Ax + B$$

Where B is a constant of integration. Since M can always be expressed in terms of x then the double integral will yield the deflection y of the beam at any section distance x from the left-hand end. The constant of integration may be determined by substituting known conditions into the equations for slope and deflection.

The product EI is termed the **flexural rigidity** of the beam.

Example 7.1

For a simply supported beam, AB, 10 m long, carrying a single concentrated load of 40 kN at the centre of its span, calculate the deflections at 2 m and 6 m from A. The flexural rigidity of the beam is 20 MNm^2.

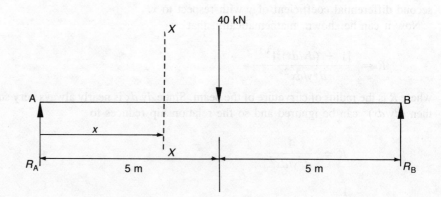

Considering the equilibrium of the external forces acting on the beam, the support reactions are:

$$R_A = 20 \text{ kN}, \quad R_B = 20 \text{ kN}$$

For the portion of the beam from A to the point of application of the load, the bending moment at xx a distance x from A is given by

$$M = 20x$$

From eqn (7.3) $\dfrac{d^2y}{dx^2} = \dfrac{M}{EI}$

substituting for M $d^2y/dx^2 = \dfrac{20x}{EI}$

Integrating with respect to x, $dy/dx = \dfrac{1}{EI} \displaystyle\int 20x\ dx + A$

$$= \frac{10x^2}{EI} + A$$

Integrating again $y = \dfrac{1}{EI} \displaystyle\int 10\ x^2\ dx + Ax + B$

$$= \frac{10x^3}{3\ EI} + Ax + B$$

The constants of integration can be found by substituting known conditions that apply in the portion of the beam under consideration, i.e. in the portion from $x = 0$ to $x = 5$. These are:

  when $x = 0$, $y = 0$ (deflection is zero at end A)

  when $x = 5$, $dy/dx = 0$ (slope is zero at centre span due to symmetry of loading)

Substituting these values in $dy/dx = \dfrac{10\ x^2}{EI} + A$

  gives $A = -\dfrac{250}{EI}$

and in $y = \dfrac{10\ x^3}{3\ EI} + Ax + B$

  gives $B = 0$

Thus the expression for the deflection becomes

$$y = \frac{1}{EI}\left\{\frac{10x^3}{3} - 250x\right\}$$

the deflection is required when $x = 2$, and in this case

$$y = \frac{1}{EI}\left\{\frac{10\cdot2^3}{3} - 250\cdot2\right\}$$

$$= -\frac{473\cdot3}{EI}$$

$$= - \frac{473 \cdot 3}{20 \times 10^6}$$

$$= 23 \cdot 67 \times 10^{-6} \text{ m}$$

$$= \underline{0 \cdot 0237 \text{ mm}}$$

The deflection at $x = 6$ m lies in the portion of the beam from the point load to B, i.e. from $x = 5$ to $x = 10$, where the bending moment expression is:

$$M = 20x - 40(x - 5)$$

or, simplifying $M = -20x + 200$

Substituting in equation (7.3) gives

$$\frac{d^2y}{dx^2} = \frac{-20x + 200}{EI}$$

Integrating $\dfrac{dy}{dx} = \dfrac{-10x^2 + 200x}{EI} + C$

and again $y = \dfrac{\dfrac{-10x^3}{3} + 100x^2}{EI} + Cx + D$

The known conditions that apply to this portion of the beam are:

when $x = 10$, $y = 0$ (deflection is zero at end)

when $x = 5$, $\dfrac{dy}{dx} = 0$ (as before slope is zero at the centre span)

Substituting these values gives

$$C = -\frac{750}{EI}, \text{ and } D = \frac{833 \cdot 3}{EI}$$

Thus the expression for the deflection in this section becomes

$$y = -\frac{10x^3}{3 \, EI} + \frac{100x^2}{EI} - \frac{750x}{EI} + \frac{833 \cdot 3}{EI}$$

The deflection when $x = 6$ m is

$$y = \frac{1}{EI} \left(-\frac{10 \times 6^3}{3} + 100 \times 6^2 - 750 \times 6 + 833 \cdot 3 \right)$$

$$y = \frac{1}{EI} (-720 + 3600 - 4500 + 833 \cdot 3)$$

$$= -\frac{786 \cdot 7}{EI}$$

$$= -\frac{786\cdot7}{20 \times 10^6}$$

$$= -\underline{0\cdot0393 \text{ mm}}$$

Example 7.2

A cantilever of length $2\cdot4$ m supports a uniformly distributed load over its entire length of 10 kN/m. If I for the cross-section of the beam is $8\cdot5 \times 10^{-6}$ m^4, and $E = 208$ GN/m^2 calculate the slope and deflection of the beam at the free end.

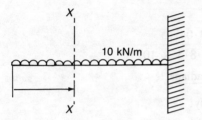

The bending moment, M, at any section xx at a distance x from the free end is given by:

$$M = 10x \cdot x/2 = 5x^2$$

from equation (7.3) $d^2y/dx^2 = \dfrac{M}{EI}$

$$= \frac{5x^2}{EI}$$

Integrating $dy/dx = \dfrac{5x^3}{3\ EI} + A$

and again $y = \dfrac{5x^4}{12\ EI} + Ax + B$

The boundary conditions that may be applied are those for any portion of the beam i.e.

$x = 2\cdot4,\quad y = 0$ (deflection zero at built-in end)
$x = 2\cdot4,\ dy/dx = 0$ (slope is zero at built-in end)

Substituting values into equations for slope and deflection gives

$$0 = \frac{5 \times 2\cdot4^3}{3\ EI} + A$$

whence $A = -\dfrac{23\cdot04}{EI}$

$$\text{and } 0 = \frac{5 \times 2 \cdot 4^4}{12 \, EI} - \frac{23 \cdot 04 \times 2 \cdot 4}{EI} + B$$

$$\text{whence } B = \frac{41 \cdot 47}{EI}$$

The slope at any section is then given by

$$dy/dx = \frac{5x^3}{3 \, EI} - \frac{23 \cdot 04}{EI}$$

at the free end, $x = 0$

$$\therefore \text{ slope at free end } = \frac{-23 \cdot 04}{208 \times 10^9 \times 8 \cdot 5 \times 10^{-6}}$$

$$= 1 \cdot 3 \times 10^{-5} \text{ radians}$$

The deflection at any section is given by:

$$y = \frac{5x^4}{12 \, EI} - \frac{23 \cdot 04x}{EI} + \frac{41 \cdot 47}{EI}$$

$$\text{when } x = 0, \quad y = \frac{41 \cdot 47}{EI}$$

$$= \frac{41 \cdot 47}{200 \times 10^9 \times 8 \cdot 5 \times 10^{-6}} = 0 \cdot 0235 \text{ mm}$$

Example 7.3

Derive expressions for the maximum slope and deflection of a simply supported beam AB of length ℓ carrying a uniformly distributed load w N/m and a point load W at its centre span.

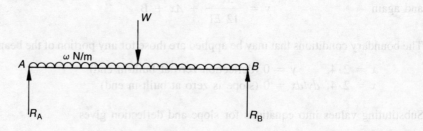

For equilibrium of the beam, $R_A = R_B = \dfrac{W + w\ell}{2}$

The expression for the bending moment at any section in the portion of the beam from A to the point of application of the point load is:

$$M = \left(\frac{W + w\ell}{2} \right) x - \frac{wx^2}{2}$$

Substituting in (7.3)

$$\frac{d^2y}{dx^2} = \frac{(W + w\ell)x - wx^2}{2 \, EI}$$

Integrating $\dfrac{dy}{dx} = \dfrac{1}{2 \, EI} \left\{ \dfrac{(W + w\ell)x^2}{2} - \dfrac{wx^3}{3} \right\} + A$

and again $y = \dfrac{1}{2 \, EI} \left\{ \dfrac{(W + w\ell)x^3}{6} - \dfrac{wx^4}{12} \right\} + Ax + B$

Substituting boundary conditions

when $x = \ell/2$, $dy/dx = 0$

hence $0 = \dfrac{1}{2 \, EI} \left\{ \dfrac{(W + w\ell)}{8} \, \ell^2 - \dfrac{w\ell^3}{24} \right\} + A$

re-arranging this gives $A = \dfrac{-w\ell^2}{16 \, EI} - \dfrac{w\ell^3}{24 \, EI}$

when $x = 0$, $y = 0$, hence $B = 0$
The general expression for slope is then

$$dy/dx = \frac{1}{2 \, EI} \left\{ \frac{(W + w\ell)}{2} x^2 - \frac{wx^3}{3} \right\} - \frac{W\ell^2}{16 \, EI} - \frac{w\ell^3}{24 \, EI}$$

The maximum slope occurs when $x = 0$ (or $x = \ell$) then

$$dy/dx = - \frac{W\ell^2}{16 \, EI} - \frac{w\ell^3}{24 \, EI} \text{ radians} \left(\text{or } \frac{W\ell^2}{16 \, EI} + \frac{w\ell^3}{24 \, EI} \right)$$

Note that, as we would expect, the slope is negative at the left-hand end of the beam and positive at the right-hand end.
   The general expression for deflection is

$$y = \frac{1}{2 \, EI} \left\{ \frac{(W + w\ell)}{6} x^3 - \frac{wx^4}{12} \right\} - \left\{ \frac{W\ell^2}{16 \, EI} + \frac{w\ell^3}{24 \, EI} \right\} x$$

The maximum deflection occurs at the centre when

$$x = \ell/2$$

then $y = \dfrac{1}{2 \, EI} \left\{ \dfrac{W + w\ell}{6} \times \dfrac{\ell^3}{8} - \dfrac{w\ell^4}{192} \right\} - \dfrac{W\ell^3}{32 \, EI} - \dfrac{w\ell^4}{48 \, EI}$

This reduces to:

$$y = -\frac{W\ell^3}{48\,EI} - \frac{5\,w\ell^4}{384\,EI}$$

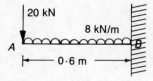

Guided solution

For the cantilever beam shown, calculate the deflection of the beam

a at the free end

b at a distance of $0 \cdot 2$ m from the free end.

Take $E = 200$ GN/m^2 and $I = 8 \cdot 8 \times 10^{-10}$ m^4.

Step 1. If the distance is calculated from the free end there will be no need to find the reactions at the wall. Write down an expression for the bending moment at any section xx, a distance x from A.

Step 2. Substitute the expression into the formula

$$d^2y/dx^2 = \frac{M}{EI}$$

and integrate twice to obtain expressions for dy/dx and y.

Step 3. Substitute suitable boundary conditions for which your expression for bending moment applies, and find values for the constants of integration

$$\left(-\frac{3 \cdot 888}{EI}, \ -\frac{1 \cdot 5804}{EI}\right)$$

Step 4. Re-write the expressions for slope and deflection to contain the calculated values of the constants of integration.

Step 5. For answer **a** substitute $x = 0$ into the slope and deflection equations and obtain a numerical value for slope and deflection by substituting for E and I

$$(y = -8 \cdot 98 \text{ mm}, \ dy/dx = 0 \cdot 221 \text{ radians})$$

Step 6. For answer **b** repeat step 5 with $x = 0 \cdot 2$

$$(y = -4 \cdot 263 \text{ mm}, \ dy/dx = 0 \cdot 0198 \text{ radians})$$

Problems 7.1

1 A simply supported beam AB carries a concentrated load of 100 kN at its mid-span. Calculate the maximum deflection of the beam if $I = 1 \cdot 7 \times 10^{-4}$ m^4, $E = 208$ GN/m^2 and the length of the beam is 6 m.

2 A cantilever beam AB, 2 m long, carries a uniformly distributed load of 20 kN/m over its entire length. Calculate the slope and deflection at its free end if $E = 207$ GN/m^2 and $I = 2 \cdot 1 \times 10^7$ mm^4.

3 A beam AB is simply supported at A and B and carries a point load of 70 kN at its mid-span, together with a uniformly distributed load of 15 kN/m over its entire length of 5 m. If the flexural rigidity, EI, of the beam is 40 MNm^2, calculate the maximum slope and deflection.

4 A cantilever beam 2 m long has a value for E of 200 GN/m^2 and $I = 8 \times 10^{-8}$ m^4. Calculate the maximum value of point load that may be applied at its extreme end, if the maximum deflection must not exceed $0\cdot1$ mm.

5 A simply supported beam of span 4 m carries two concentrated loads each of 50 kN at 1 m and 3 m from one end. Calculate the slope and deflection under each load, and at the mid-span.

7.3   Standard cases for beam slope and deflection

In many instances it will be the maximum values of slope and deflection which will most interest the engineer. In the cases of the most common loading and support systems expressions for the maximum slope and deflection have been derived and are readily available for the engineer to use. Some of the standard cases are listed below in the table, (Fig. 7.3).

	Maximum slope	Maximum deflection
W ↓, ℓ (cantilever, point load)	$\dfrac{W\ell^2}{2EI}$ (at free end)	$\dfrac{W\ell^3}{3EI}$ (at free end)
ω/unit length, ℓ (cantilever, UDL)	$\dfrac{\omega\ell^3}{6EI}$ (at free end)	$\dfrac{\omega\ell^4}{8EI}$ (at free end)
W at mid-span, $\ell/2 + \ell/2$	$\dfrac{W\ell^2}{16EI}$ (at each end)	$\dfrac{W\ell^3}{48EI}$ (at centre)
ω/unit length, ℓ (simply supported, UDL)	$\dfrac{\omega\ell^3}{24EI}$ (at each end)	$\dfrac{5\omega\ell^4}{384EI}$ (at centre)
W, $a + b$, ℓ		Deflection under the load = $\dfrac{Wa^2b^2}{eEI\ell}$

Fig. 7.3

7.4   The principle of superposition

We have assumed that the deflection at a point of a beam is very small, and indeed the examples to date have confirmed this assumption. The effect of additional loads will be to cause further deflection, and if we assume the beam to be virtually straight even after the initial loading then this extra deflection can be attributed solely to the additional loads. This means that the deflection due to several loads can be assumed to be the sum of the individual deflections due to the loads taken separately. This method of calculating beam deflection is known as the principle of superposition.

Example 7.4

Consider again the beam in example 7.3. The load system can be considered as two separate systems for which the deflections can be found using the standard cases, listed in section 7.3.

Thus, the given system can be regarded as

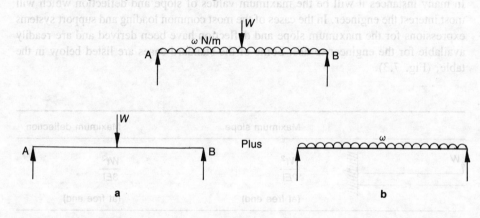

Using the standard cases

The maximum deflection for **a** $= + \dfrac{W\ell^3}{48\ EI}$

and the maximum deflection for **b** $= - \dfrac{5\ w\ell^4}{384\ EI}$

Thus the total deflection at the centre (the maximum deflection)

$$= - \frac{W\ell^3}{48\ EI} - \frac{5\ w\ell^4}{384\ EI}$$

which is the same result as before.

Similarly, the maximum slope for **a** $= - \dfrac{W\ell^2}{16\ EI}$

and the maximum slope for **b** $= - \dfrac{w\ell^3}{24\ EI}$

(both at the left-hand end). Thus the total slope at the left-hand end of the beam is

$$-\frac{W\ell^2}{16\,EI} - \frac{w\ell^3}{24\,EI}$$

again, the same result as before.

Example 7.5

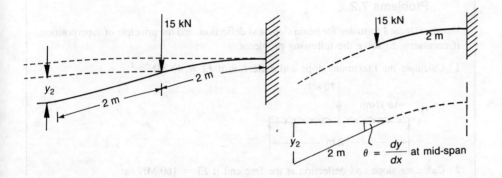

$\theta = \dfrac{dy}{dx}$ at mid-span

A cantilever beam, 4 m long carries a concentrated load of 15 kN at a point 2 m from the free end. If EI for the beam is 25 MNm^2 calculate the deflection at the free end. The deflection y_1 that takes place under the point load may be calculated by assuming that the 15 kN acts at the end of a cantilever beam 2 m long.

The deflection is then given by

$$y_1 = \frac{W\ell^3}{3\,EI}$$

$$= \frac{15 \times 10^3 \times 2^3}{3 \times 25 \times 10^6}$$

$$= 0 \cdot 0016 \text{ m}$$

$$= \underline{1 \cdot 6 \text{ mm}}$$

and the slope by $dy/dx = \dfrac{W\ell^2}{2\,EI} = \dfrac{15 \times 10^3 \times 2^2}{2 \times 25 \times 10^6} = 0 \cdot 0012$ radian

Since there is no load acting on the beam between the 15 kN load and the free end, then this portion of the beam will not be bent, but is simply a straight line of slope equal to the slope under the 15 kN load. The deflection y_2 is then found by simple trigonometry.

Thus, $y_2 = 2 \sin \theta$
              $= 2\theta$

since $\sin \theta = \theta$ in radians for very small angles.

Thus $y_2 = 2 \times 0 \cdot 0012$

$$= 0 \cdot 0024 \text{ m}$$
$$= 2 \cdot 4 \text{ mm}$$

The total deflection at the end of the cantilever

$$= y_1 + y_2 = 1 \cdot 6 \text{ mm} + 2 \cdot 4 \text{ mm}$$
$$= \underline{4 \text{ mm}}$$

Problems 7.2

Use the standard formulae for beam slope and deflection, and the principle of superposition if necessary, to solve the following problems.

1 Calculate the maximum slope and deflection if $EI = 40 \text{ MN m}^2$.

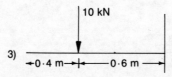

2 Calculate slope and deflection at the free end if $EI = 160 \text{ MN m}^2$.

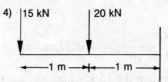

3 Calculate slope and deflection under the load, and the deflection at the free end. $EI = 20 \text{ MN m}^2$.

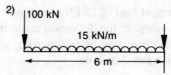

4 Calculate the deflection at the free end. $EI = 20 \text{ MN m}^2$.

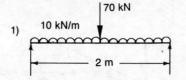

5 Calculate the deflection under the load. $EI = 21 \text{ MN m}^2$.

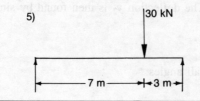

7.5   Macaulay's method for beam deflection

It is important to appreciate that the method used so far to calculate beam deflections depends on the calculation of a bending moment expression *for the portion of the beam in which the deflection is required*. Clearly for a beam with a number of applied loads the same number of applications of the basic formula as there are loads will be required if a deflection in every portion of the beam is wanted. In addition only boundary conditions that apply to the portion of the beam under consideration can be used and often there is insufficient information to give a solution. It is for these reasons that Macaulay's method for beam deflections was devised.

The method depends on a single expression for bending moment which applies to the whole beam and so enables boundary conditions for any part of the beam to be used. A condition for use of the method is that certain terms are ignored for particular values of x, the distance along the beam. The method is best explained by reference to an example.

Consider the simply supported beam shown subjected to three point loads, W_1, W_2 and W_3 at distances a, b and c respectively from the left-hand end. Let the support reactions be R_1 and R_2 (Fig. 7.4).

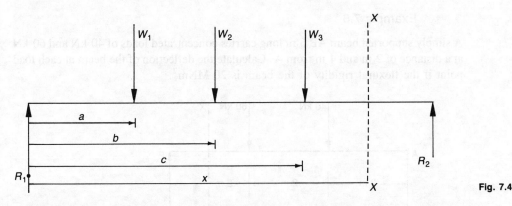

Fig. 7.4

An expression for bending moment is written for a section xx a distance x from the left-hand end, in the *final* portion of the beam, i.e. the expression will include *all* of the applied loads.

Thus $M = R_1 x - W_1[x - a] - W_2[x - b] - W_3[x - c]$

since $d^2y/dx^2 = \dfrac{M}{EI}$

then $d^2y/dx^2 = \dfrac{1}{EI}(R_1 x - W_1[x - a] - W_2[x - b] - W_3[x - c])$

Note the use of square brackets [ ] in the bending moment expression, these brackets are integrated as a whole.

Thus, $dy/dx = \dfrac{1}{EI}\left(R_1\dfrac{x^2}{2} - W_1\dfrac{[x - a]^2}{2} - W_2\dfrac{[x - b]^2}{2}\right)$

$$- W_3 \frac{[x - c]^2}{2} + A \bigg)$$

and $\quad y = \dfrac{1}{EI} \bigg(\dfrac{R_1 x^3}{6} - W_1 \dfrac{[x - a]^3}{6} - W_2 \dfrac{[x - b]^3}{6}$

$$- W_3 \frac{[x - c]^3}{6} + Ax + B \bigg)$$

When substituting boundary conditions those that apply to any part of the beam can be used since the bending moment expression applies to the whole beam. Any square brackets which are negative are ignored. Thus, in this case:

when $\quad x = 0$, $y = 0$ $\therefore$ $B = 0$
and when $\quad x = \ell$, $y = 0$ which will give A.

The slope and deflection can then be found at any point on the beam, *providing that all square brackets which come out negative are ignored.*

Example 7.6

A simply supported beam AB 6 m long carries concentrated loads of 40 kN and 60 kN at a distance of 2 m and 4 m from A. Calculate the deflection of the beam at each load point if the flexural rigidity of the beam is 20 MNm^2.

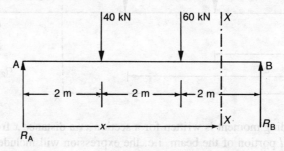

From equilibrium of moments $R_A = 46\frac{2}{3}$ kN, $R_B = 53\frac{1}{3}$ kN. The expression for bending moment M in the final portion of the beam is

$$M = R_A x - 40[x - 2] - 60[x - 4]$$

Thus $\qquad d^2 y/dx^2 = \dfrac{1}{EI} (R_A x - 40[x - 2] - 60[x - 4])$

Integrating $\quad dy/dx = \dfrac{1}{EI} \bigg(R_A \dfrac{x^2}{2} - 20[x - 2]^2 - 30[x - 4]^2 \bigg) + A$

and again $\qquad y = \dfrac{1}{EI} \bigg(R_A \dfrac{x^3}{6} - \dfrac{20}{3} [x - 2]^3 - 10[x - 4]^3 \bigg) + Ax + B$

Boundary conditions that apply to the whole beam are:

when $x = 0$, $y = 0$
when $x = 6$, $y = 0$

Note that these boundary conditions are the only ones we know anything about. Since the beam is not symmetrically loaded the maximum deflection, and hence zero slope, is not at the mid-span.

In fact there is no information available to determine the position of zero slope, and so without Macanlay's method a solution would not be possible.

Substituting the known boundary conditions into the expression for deflection, and remembering that terms in square brackets that are negative can be ignored gives:

$$B = 0$$

and

$$0 = \frac{1}{EI}\left(46\tfrac{2}{3} \times \frac{6^3}{6} - \frac{20}{3} \times 4^3 - 10 \times 2^3\right) + 6A$$

whence $A = -\dfrac{195 \cdot 58}{EI}$

Thus the general expression for deflection is:

$$y = \frac{1}{EI}\left(46 \cdot 67\, \frac{x^3}{6} - \frac{20}{3}[x-2]^3 - 10[x-4]^3\right) - \frac{195 \cdot 58}{EI}\, x$$

when $x = 2$ m, $y = \dfrac{1}{EI}\left(\dfrac{46 \cdot 67 \times 2^3}{6} - 195 \cdot 58 \times 2\right)$

(remembering that negative terms in square brackets are ignored)

$$y = -\frac{328 \cdot 93}{EI}$$

$$= -\frac{328 \cdot 93}{20 \times 10^6} = -0 \cdot 0164 \text{ mm}$$

when $x = 4$ m, $y = \dfrac{1}{EI}\left(\dfrac{46 \cdot 67 \times 4^3}{6} - \dfrac{20}{3} \times 2^3 - 195 \cdot 58 \times 4\right)$

$$= -\frac{337 \cdot 85}{EI}$$

$$= -\frac{337 \cdot 85}{20 \times 10^6} = -0 \cdot 0169 \text{ mm}$$

Example 7.7

For the cantilever beam AB shown, calculate the slope and deflection at the free end. $EI = 30$ MN m^2. Since the free end A is the origin of co-ordinates there is no need to calculate the reactions at the built-in end.

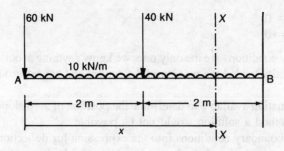

The expression for the bending moment in the final portion of the beam is:

$$M = -60x - 40[x - 2] - 10x \, x/2$$

Thus $d^2y/dx^2 = \dfrac{1}{EI} (-60x - 40[x - 2] - 5x^2)$

$$dy/dx = \frac{1}{EI} \left(-30x^2 - 20[x - 2]^2 - \frac{5x^3}{3} \right) + A$$

$$y = \frac{1}{EI} \left(-10x^3 - \frac{20}{3} [x - 2]^3 - \frac{5x^4}{12} \right) + Ax + B$$

the boundary conditions are, when $x = 4$, $y = 0$
and when $x = 4$, $dy/dx = 0$

Thus $0 = \dfrac{1}{EI} \left(-30. \, 4^2 - 20 \, [2]^2 - 5. \, \dfrac{4^3}{3} \right) + A$

whence $A = \dfrac{666 \cdot 67}{EI}$

and $0 = \dfrac{1}{EI} \left(-10 \times 4^3 - \dfrac{20}{3} [2]^3 - \dfrac{5 \times 4^4}{12} + 666 \cdot 67 \right) + B$

whence $B = -\dfrac{133 \cdot 33}{EI}$

The slope when $x = 0$ is then given by $dy/dx = \dfrac{666 \cdot 67}{EI} = 0 \cdot 0222$ rads

The deflection when $x = 0$ is given by $y = -\dfrac{133 \cdot 33}{EI}$

$$= \underline{0 \cdot 0044 \text{ mm}}$$

Example 7.8

A simply supported beam AB, 10 m long carries concentrated loads of 200 kN and 300 kN at a distance of 2 m and 6 m from A respectively. Calculate the maximum deflections of the beam if $EI = 80$ MNm^2.

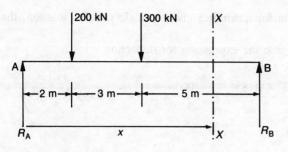

The support reactions are $R_A = 310$ kN and $R_B = 190$ kN. The bending moment in the final portion of the beam is given by:

$$M = 310x - 200[x - 2] - 300[x - 5]$$

Then, $d^2y/dx^2 = \dfrac{1}{EI} (310x - 200[x - 2] - 300[x - 5])$

$$dy/dx = \frac{1}{EI} (155x^2 - 100[x - 2]^2 - 150[x - 5]^2) + A$$

$$y = \frac{1}{EI} \left(\frac{155x^3}{3} - \frac{100}{3} [x - 2]^3 - 50[x - 5]^3 \right) + Ax + B$$

when $x = 0$, $y = 0$ $B = 0$ (neglecting negative square brackets)

when $x = 10$, $y = 0$ $0 = \dfrac{1}{EI} \left(\dfrac{155 \times 10^3}{3} - \dfrac{100}{3} \cdot 8^3 - 50 \times 5^3 \right) + 10A$

whence $A = -\dfrac{2835}{EI}$

The expressions for slope and deflection are thus

$$dy/dx = \frac{1}{EI} (155x^2 - 100[x - 2]^2 - 150[x - 5]^2 - 2835)$$

and $y = \dfrac{1}{EI} \left(\dfrac{155x^3}{3} - \dfrac{100}{3} [x - 2]^3 - 50[x - 5]^3 - 2835x \right)$

The maximum deflection occurs when the slope is zero. A difficulty arises in deciding which of the terms in square brackets in the expression for slope will be negative and so can be ignored. It is safe to assume that the given load distribution will produce a maximum deflection to the left of centre, and so the term $[x - 5]$ can be ignored.

Thus for maximum deflection

$$0 = 155x^2 - 100[x - 2]^2 - 2835$$

simplifying this gives:

$$11x^2 + 80x - 647 = 0$$

Solving using the formula for quadratics gives a single positive solution, thus $x = 4 \cdot 85$ m.

Substituting this value into the expression for deflection

$$y_{max} = \frac{1}{EI} \left(\frac{155 \times 4 \cdot 85^3}{3} - \frac{100}{3} \times 2 \cdot 85^3 - 2835 \times 4 \cdot 85 \right)$$

$$= - \frac{8627}{EI}$$

$$= - \frac{8627}{80 \times 10^6} \times 10^3 \text{ mm}$$

$$= \underline{0 \cdot 108 \text{ mm}}$$

Thus the maximum deflection of the beam is $0 \cdot 108$ mm, and it occurs at a point $4 \cdot 85$ m from A.

In the last worked example the approximate position of the maximum deflection was reasonably easy to decide. In more complex cases it is often necessary to hazard a guess at the portion of the beam which has the maximum deflection, and if the calculated value of x is indeed in that portion, the guess is correct. If the value of x is outside the chosen portion of the beam it is then necessary to repeat the calculation with a changed portion.

It is also important to point out that in the case of cantilevers, or beams with overhangs, the maximum deflection may not occur at zero slope, but at the free ends, and in these cases a separate calculation is required.

Example 7.9

Calculate the position and value of the maximum deflection for the beam shown if $EI = 100$ MNm^2.

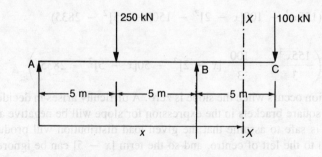

The reactions are: $R_A = 85$ kN $R_B = 265$ kN.

The bending moment expression for the last portion of the beam is

$$M = 85x + 265[x - 10] - 250[x - 5]$$

Thus, $d^2y/dx^2 = \dfrac{1}{EI}(85x + 265[x - 10] - 250[x - 5])$

$$dy/dx = \frac{1}{EI}\left(\frac{85x^2}{2} + \frac{265}{2}[x - 10]^2 - \frac{250}{2}[x - 5]^2\right) + A$$

$$y = \frac{1}{EI}\left(\frac{85x^3}{6} + \frac{265}{6}[x - 10]^3 - \frac{250}{6}[x - 5]^3\right) + Ax + B$$

when $x = 0$, $y = 0$ $B = 0$

when $x = 10$, $y = 0$ $0 = \dfrac{1}{EI}\left(\dfrac{85 \times 10^3}{6} + \dfrac{265}{6}[0]^3 - \dfrac{250 \times 5^3}{6} + 10A\right)$

whence $A = -\dfrac{895\cdot8}{EI}$

The expressions for slope and deflection then become

$$dy/dx = \frac{1}{EI}\left(\frac{85x^2}{2} + \frac{265}{2}[x - 10]^2 - \frac{250}{2}[x - 5]^2 - 895\cdot8\right)$$

and $y = \dfrac{1}{EI}\left(\dfrac{85x^3}{6} + \dfrac{265}{6}[x - 10]^3 - \dfrac{250}{6}[x - 5]^3 - 895\cdot8x\right)$

Guessing that the maximum deflection occurs in the portion of the beam from $x = 0$ to $x = 5$ then $dy/dx = 0$, when

$$\frac{85x^2}{2} - 895\cdot8 = 0$$

i.e. $x^2 = \dfrac{895\cdot8 \times 2}{85}$ giving $\underline{x = 4\cdot59 \text{ m}}$

The local maximum deflection is then:

$$y_{max} = \frac{1}{EI}\left(\frac{85 \times 4\cdot59^3}{6} - 895\cdot8 \times 4\cdot59\right)$$

$$= -\frac{2741\cdot8}{EI}$$

$$= -\frac{2741\cdot8}{100 \times 10^6} \times 10^3 \text{ mm} = \underline{0\cdot0274 \text{ mm}}$$

The deflection at the overhang when $x = 14$ m is given by:

$$y = \frac{1}{EI}\left(\frac{85 \times 14^3}{6} + \frac{265}{6}.4^3 - \frac{250}{6}.9^3 - 895\cdot8 \times 14\right)$$

$$= -\frac{1216 \cdot 2}{EI}$$

$$= -\frac{1216 \cdot 2}{100 \times 10^6} \times 10^3 \text{ mm}$$

$$= -\underline{0 \cdot 01216 \text{ mm}}$$

Thus the maximum deflection of $0 \cdot 0274$ mm occurs at a point $4 \cdot 59$ m from A.

7.6   Macaulay's method for uniformly distributed loads over part of a beam's span

When a uniformly distributed load does not continue to the right-hand end of the beam then the following procedure must be used before writing down the Macaulay expression. Firstly the UDL is continued to the right-hand end, and secondly an equal, but opposite loading is introduced to counteract the addition. For example, the beam in Fig. 7.5a is modified to the equivalent system in Fig. 7.5b.

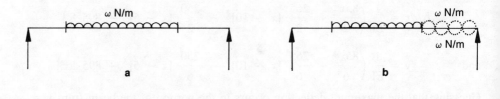

Fig. 7.5

a   b

Example 7.10

A beam ABC is simply supported at A and B 2 m apart, and carries a uniformly distributed load of 50 kN/m between A and B, together with a concentrated load of 25 kN on an overhang at C, $0 \cdot 5$ m from B. Calculate the maximum deflection of the beam, given that $EI = 21$ MN m^2.

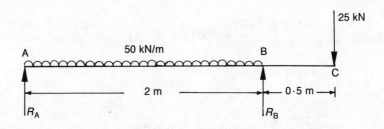

From the equilibrium of moments $R_A = 43 \cdot 75$ kN

and $R_B = 81 \cdot 25$ kN

Since the UDL does not extend to the extreme right-hand end of the beam then the loading is modified thus:

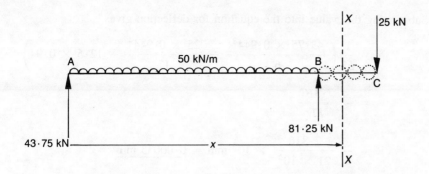

The expression for bending moment in the final portion of the beam is

$$M = 43 \cdot 75x + 81 \cdot 25[x - 2] - 50x \cdot \frac{x}{2} + 50[x - 2] \frac{[x - 2]}{2}$$

where the term $50[x - 2] \dfrac{[x - 2]}{2}$ represents the moment of the *added* UDL.

Thus $d^2y/dx^2 = \dfrac{1}{EI} (43 \cdot 75x + 81 \cdot 25[x - 2] - 25x^2 + 25[x - 2]^2)$

$$dy/dx = \frac{1}{EI} \left(\frac{43 \cdot 75x^2}{2} + \frac{81 \cdot 25}{2} [x - 2]^2 - \frac{25x^3}{3} + \frac{25}{3} [x - 2]^3 \right) + A$$

$$y = \frac{1}{EI} \left(\frac{43 \cdot 75x^3}{6} + \frac{81 \cdot 25}{6} [x-2]^3 - \frac{25x^4}{12} + \frac{25}{12} [x-2]^4 \right) + Ax + B$$

when $x = 0$, $y = 0$ $\therefore B = 0$

when $x = 2$, $y = 0$ $\therefore 0 = \dfrac{43 \cdot 75 \times 2^3}{6\ EI} - \dfrac{25 \times 2^4}{12\ EI} + 2A$

whence $A = -\dfrac{12 \cdot 5}{EI}$

The maximum deflection between the supports occurs when

  $dy/dx = 0$

i.e. $\dfrac{43 \cdot 75x^2}{2} - \dfrac{25x^3}{3} - 12 \cdot 5 = 0$ ignoring terms in square brackets.

Multiplying through by 6 and re-arranging gives

  $50x^3 - 131 \cdot 25x^2 + 75 = 0$

A solution to this equation can be found by using Newton's Method or by trial and error.
  In this case $x = 0 \cdot 944$ is the approximate solution in the required range.

Substituting this value into the equation for deflection gives

$$y = \frac{1}{EI} \left(\frac{43 \cdot 75 \times 0 \cdot 944^3}{6} - \frac{25 \times 0 \cdot 944^4}{12} - 12 \cdot 5 \times 0 \cdot 94 \right)$$

$$= -\frac{7 \cdot 316}{EI}$$

$$= -\frac{7 \cdot 316}{21 \times 10^6} \times 10^3 \text{ mm} = \underline{0 \cdot 00035 \text{ mm}}$$

The deflection at the free end when $x = 2 \cdot 5$ is given by

$$y = \frac{1}{EI} \left(\frac{43 \cdot 75 \times 2 \cdot 5^3}{6} + \frac{81 \cdot 25}{6} \times 0 \cdot 5^3 - \frac{25}{12} \times 2 \cdot 5^4 + \frac{25}{12} \right.$$

$$\left. \times 0 \cdot 5^4 - 12 \cdot 5 \times 2 \cdot 5 \right)$$

$$= \frac{3 \cdot 09}{EI}$$

$$= \frac{3 \cdot 09}{21 \times 10^6} \times 10^3 \text{ mm} = \underline{0 \cdot 0015 \text{ mm}}$$

Hence the maximum deflection of the beam is $0 \cdot 00035$ mm at a point $0 \cdot 944$ m from A.

Guided solution

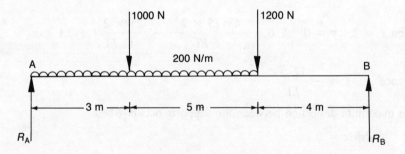

For the beam AB shown calculate the deflections under the concentrated loads. $EI = 25 \text{ MNm}^2$.

Step 1.   Calculate the value of the support reactions.

$$(R_A = 2216 \cdot 7 \text{ N}, \quad R_B = 1583 \cdot 3 \text{ N})$$

Step 2.   Extend the UDL to the end of the beam B and counteract this by equal but opposite loading on the underside of the beam.

Step 3. Write down an expression for bending moment in the final portion, measured from A, of the modified beam.

Step 4. Substitute the expression for bending moment into the formula for d^2y/dx^2, and integrate twice, remembering that square brackets are integrated as a whole.

Step 5. Substitute boundary conditions and evaluate the constants of integration. Remember that square brackets which are negative must be ignored.

$$\left(0; \; -\frac{22748}{EI}\right)$$

Step 6. Substitute $x = 3$ and $x = 8$ into the completed expression for deflection to obtain the required values

$$(-2\cdot358 \text{ mm at } x = 3, \; -1\cdot91 \text{ mm at } x = 8)$$

Problems 7.3

1 A beam ABCDE is simply supported at A and E and carries concentrated loads of 10 kN, 20 kN and 30 kN at B, C and D respectively. If AB = 1 m, and BC = 2 m, CD = 2 m and DE = 2 m calculate the deflection at each load point. $EI = 25$ MN m^2.

2 A cantilever beam 3 m long carries a uniformly distributed load of 200 N/m over its entire length, together with concentrated loads of 400 N at its free end, and 600 N at a point $1\cdot5$ m from the free end. If $EI = 20$ MNm^2 calculate the slope and deflection at the free end.

3 A simply supported beam AB of length 10 m carries point loads of 20 kN and 30 kN at points 3 m and 5 m from A respectively. If $EI = 40$ MN m^2 calculate the position and value of the maximum deflection.

4 A beam ABC, 8 m long, is simply supported at A and B. The beam carries concentrated loads of 50 kN at a point 2 m from A, and 40 kN at C which is on an overhang 2 m from B. $EI = 30$ MNm^2. Calculate the position and value of the maximum deflection.

5 A beam ABCD consists of a hollow circular tube of outside diameter 150 mm and inside diameter 125 mm, simply supported at B and D. AB = 1 m, BC = 2 m and CD = $1\cdot5$ m. The beam carries a UDL of 40 kN/m over AC, and a concentrated load of 100 kN at C. Determine the slope of the beam at B and the deflection at A and C. $E = 200$ GN/m^2.

6 A beam ABC is 12 m long, and is simply supported at A and B. The beam carries a UDL of 20 kN/m over AB, and a point load of 60 kN at C, which is 3 m from B. If $EI = 42$ MNm^2 calculate the maximum deflection of the beam.

7 A beam ABCDE is simply supported at B and D. AB = 2 m, BC = 3 m, CD = 2 m and DE = 1 m. The beam carries a UDL of 10 kN/m over AC, and concentrated loads of 40 kN and 20 kN at C and E respectively. If $EI = 15$ MNm^2 calculate the position and magnitude of the maximum deflection.

8 Friction applications

The contents of this chapter will enable you to demonstrate an understanding of the important role of friction in power transmission by belts and clutches, and the power losses due to friction that occur in plain bearings.

8.1 Belt drives

Belt drives use the friction that exists between the pulley and belt to transmit a torque. The belt is stretched between two pulleys with an initial tension, T_0. The driving pulley, i.e. the one supplying the torque, transmits the power via the friction between the contact surfaces to the driven pulley, the one receiving the torque, again utilising the frictional forces between the contact surfaces of belt and pulley. The transmitted torque is solely due to the difference in tensions that exists in the belt during operation, there being a 'tight' side and a 'slack' side, (Fig. 8.1).

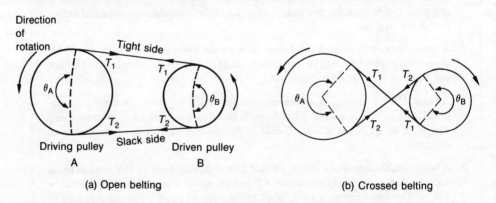

Fig. 8.1   (a) Open belting   (b) Crossed belting

Belt drives have several advantages and some disadvantages;

(i)   There is an element of safety in a belt-drive system in that shock loadings or sudden torques may be absorbed by slipping of the belt.
(ii) Different sizes of pulleys may be used to give a gearing effect. Even variable gearing may be achieved by using a system of conical pulleys.
(iii) The driven pulley in the open system, (Fig. 8.1a) rotates in the same direction as the driving pulley. This is not so easily achieved in gear mechanisms without additional gears. Counter rotation in belt drives can be achieved by cross belting, (Fig. 8.1b).

(iv) Belt materials are stretchable so that in operation a belt may extend and alter the relationships that exist between tensions, speeds and torques. If these relationships are important (e.g. in cam drives) then toothed belts which afford more positive location may be used.

(v) Hysteresis effects caused by the flexing of the belt may result in a loss of energy and a consequent reduction in the power transmitted. This may be particularly evident with a worn belt.

Consider the open belt-drive system shown in Fig. 8.1a. The driving pulley A causes a difference in the tight and slack side tensions in the belt of $T_1 - T_2$. This difference in tensions is possible because of the friction that exists between belt and pulley along the contact length of the circumference of the pulley. This contact length subtends an angle θ_A at the centre of the pulley called the **angle of lap**. The difference in tensions is then transmitted to the driven pulley B causing a torque which rotates the pulley. The angle of lap of the belt on pulley B, θ_B, will be different from that on pulley A unless the pulleys are of the same diameter. The analysis of belt drives makes the important assumption that the limiting friction case applies, i.e. that the belt is on the point of slipping over the pulley. This condition can apply not only at speed, but also when the pulleys are at rest at the instant when rotation is about to commence.

8.2   Flat belts

The relationship that exists between the tight and slack side tensions, T_1 and T_2 may be determined by considering an element of the belt of length δx which subtends an angle $\delta\theta$ at the centre of the pulley whose radius is r, (Fig. 8.2). In the limiting friction case let the difference in tensions over this element of the belt be δT, that is if the tension on one side of the element is T, the tension on the other is $T + \delta T$.

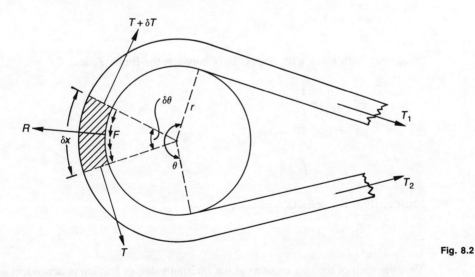

Fig. 8.2

There will also be a radial reaction force R acting on the element and supplied by the pulley, and a limiting frictional force F which will oppose the tendency for the belt to slip. In this limiting condition $F = \mu R$ where μ is the coefficient of dry friction.

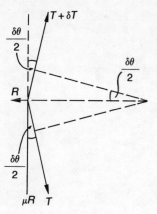

Fig. 8.3

Using the simplified diagram of Fig. 8.3 these forces may be resolved in mutually perpendicular directions for equilibrium of the element.

Resolving tangentially, and since $\delta\theta$ is very small $\cos \delta\theta/2 \simeq 1$.

$$T + \mu R \simeq T + \delta T$$

or

$$\mu R \simeq \delta T \tag{8.1}$$

Resolving radially

$$R = T \sin \frac{\delta\theta}{2} + (T + \delta T) \sin \frac{\delta\theta}{2}$$

since $\delta\theta$ is very small, $\sin \dfrac{\delta\theta}{2} \simeq \dfrac{\delta\theta}{2}$ in radians

$$\therefore R = T \frac{\delta\theta}{2} + (T + \delta T)\frac{\delta\theta}{2}$$

neglecting the product of very small quantities this reduces to

$$R \simeq T\delta\theta \tag{8.2}$$

Substituting equation (8.2) into (8.1) gives

$$\mu T\delta\theta \simeq \delta T$$

or

$$\mu\frac{\delta\theta}{\delta T} \simeq \frac{1}{T}$$

In the limit, as $\delta\theta \to 0$ this becomes

$$\mu\frac{d\theta}{dt} = \frac{1}{T}$$

Integrating both sides with respect to T between the limits T_1 and T_2

$$\mu\theta = \int_{T_2}^{T_1} \frac{1}{T}\, dT$$

or

$$\mu\theta = \log_e T_1 - \log_e T_2$$

$$= \log_e \left(\frac{T_1}{T_2}\right)$$

whence, from the definition of a $\log_e$

$$\frac{T_1}{T_2} = e^{\mu\theta} \tag{8.3}$$

Thus the ratio of the belt tensions at the limiting value of friction is dependent on the coefficient of dry friction and the angle of lap. It is evident that when one pulley is smaller than the other, as in Fig. 8.1, then the smaller angle of lap, that of the smaller pulley, should be used in determining the ratio of belt tensions to just avoid slip. Note that for crossed belts the angle of lap is the same regardless of size of pulley.

8.3   V-belts

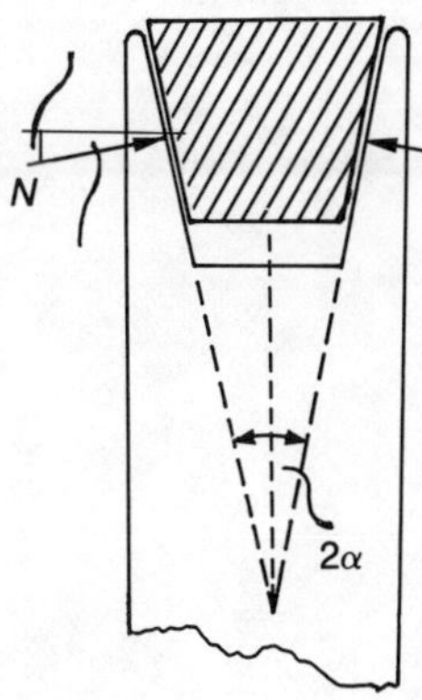

In V-belt drives the specially tapered section of the V-belt is located in V-grooves in the circumference of the pulley. This not only provides more positive location of the belt, but also increases the normal reaction between the contact surfaces of the belt and the pulley. This latter effect allows for a greater limiting belt tension ratio for a given value of μ and θ as we shall now see. Consider the cross-section of a V-belt in its groove as in Fig. 8.4. The radial reaction force is provided by the radial components of the normal reactions, N, acting on the sides of the V, that is $2N \sin \alpha$, where α is half of the V-angle. The tangential frictional force will now be $2 \mu N$.

It is straightforward to repeat the analysis used for flat belts with these new values to give the modified result

$$\frac{T_1}{T_2} = e^{\mu \frac{\theta}{\sin \alpha}} \tag{8.4}$$

Fig. 8.4

This result shows that in general a V-belt has a ratio of belt tensions equivalent to a flat belt whose coefficient of friction has been raised from μ to $\mu/\sin \alpha$.

8.4   Power transmission by belts

As stated in Section 8.1 the driving pulley causes a difference in tensions in the belt which is transferred to the driven pulley. The driven pulley thus experiences a torque which is caused by the difference in tensions in the belt.

Referring once again to Fig. 8.1:

$$\text{torque on driving pulley A} = (T_1 - T_2)r_A$$
$$\text{and, torque on driven pulley B} = (T_1 - T_2)r_B$$

where r_A and r_B are the radii of the respective pulleys.

Now, power transmitted = torque $\times$ angular velocity
$$= (T_1 - T_2)r_A\omega_A \text{ for pulley A}$$
$$= (T_1 - T_2)r_B\omega_B \text{ for pulley B}.$$

But $\omega_A = \dfrac{v}{r_A}$ and $\omega_B = \dfrac{v}{r_B}$ where v is the circumferential speed of each pulley which is equal to the belt speed, so that;

$$\text{power transmitted} = (T_1 - T_2)v \text{ watts} \tag{8.5}$$

Using the limiting condition that $\dfrac{T_1}{T_2} = e^{\mu\theta}$ (for a flat belt) then the maximum power that can be transmitted is given by

$$\text{power} = T_1 \left(1 - \frac{1}{e^{\mu\theta}}\right) v \tag{8.6}$$

A further relationship between the tight and slack tensions and the initial tension of the belt is often assumed:

i.e. $$T_o = \frac{T_1 + T_2}{2} \qquad (8.7)$$

This relationship, which states that the initial tension is the mean of the tight- and slack-side tensions, is valid providing that the belt stretch during operation is minimal.

Example 8.1

A machine fitted with a pulley of diameter 250 mm is driven by a motor fitted with a pulley of 150 mm diameter using an open flat-belt system. The centre distance of the pulleys, which are in the same plane, is 0·5 m and the speed of rotation of the motor is 1500 rev/min. If the maximum belt tension permitted is 890 N calculate the

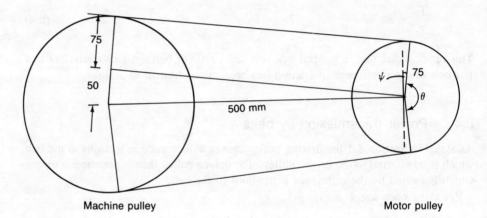

Machine pulley   Motor pulley

maximum power which can be transmitted. Take $\mu = 0·25$. The maximum power transmitted will depend on the angle of lap, θ, of the smaller pulley.

From the diagram it can be seen that

$$\sin \psi = \frac{50}{500} = \frac{1}{10}$$

or $\psi = 5° \ 44·4'$

and $\theta = 180 - 2\psi$

$= 180 - 2 \times 5° \ 44·4'$

$= 168·52°$

$= 2·94$ radians

The maximum belt tension is $T_1 = 890$ newtons.

The belt speed, $\nu =$ angular velocity of motor $\times$ radius of motor pulley

$$= \frac{2\pi \times 1500}{60} \times 0·075$$

$= 11·78$ m/s

From equation (8.6) the maximum power transmitted is given by:

$$\text{maximum power} = T_1 \left(1 - \frac{1}{e^{\mu\theta}}\right) v$$

$$= 890 \left(1 - \frac{1}{e^{0\cdot25 \times 2\cdot94}}\right) 11\cdot78$$

$$= 890 \times 0\cdot52 \times 11\cdot78$$
$$= 5452 \text{ watts}$$
$$= \underline{5\cdot45 \text{ kW}}$$

Example 8.2

A belt drive consists of 4 V-belts which connect a driving pulley of effective diameter 150 mm to a driven pulley of effective diameter 300 mm. The angle of lap of the belts on the driving pulley is 160° and the cross-sectional area of each belt is 10^{-4} m². If the maximum stress in the belts is not to exceed 3 MN/m² and $\mu = 0\cdot2$ calculate the power which can be transmitted at a belt speed of 25 m/s, and the torque that is applied to each pulley. The pulley groove angles are 40°.

The angle of lap $= 160° = 2\cdot79$ radians

The ratio of belt tensions, $\dfrac{T_1}{T_2} = e^{\frac{\mu\theta}{\sin \alpha}}$

$$= e^{\frac{0\cdot2 \times 2\cdot79}{\sin 20°}}$$

$$= 5\cdot11$$

The maximum permissible tension, T_1 = maximum permissible stress × area

$$= 3 \times 10^6 \times 10^{-4}$$
$$= 300 \text{ N}$$

Hence $T_2 \qquad\qquad = \dfrac{300}{5\cdot11} = 58\cdot7 \text{ newtons}$

The power that can be transmitted by each belt is then given by

$$\text{power} = (T_1 - T_2)v$$
$$= (300 - 58\cdot7)25$$
$$= \underline{6\cdot03 \text{ kW}}$$

Four such belts are then capable of transmitting $4 \times 6\cdot03 \text{ kW} = \underline{24\cdot1 \text{ kW}}$

Torque applied to driving pulley $= (T_1 - T_2)r$
$$= (300 - 58\cdot7) \times 0\cdot15$$
$$= \underline{36\cdot2 \text{ Nm}}$$

Similarly the torque applied to driven pulley

$$= (300 - 58 \cdot 7) \times 0 \cdot 3$$
$$= \underline{72 \cdot 4 \text{ Nm}}$$

Problems 8.1

1 A belt drive is used to transmit power to a machine pulley. The belt speed required is 2 m/s and the maximum allowable belt tension is 300 N. The angle of lap of the belt over the pulley is 140° and the coefficient of friction between the belt and pulley is 0·25. Calculate the maximum power that can be transmitted using
   a a flat belt
   b a V-belt with a groove angle of 40°.

2 **i** An open flat belt transmits 10 kW from a pulley 0·5 m diameter rotating at 500 rev/min to a pulley 0·3 m diameter. The pulley centres are 1·5 m apart and the coefficient of friction between the belt and pulleys is 0·3. Determine the maximum tension in the belt.
   ii If the belt is now crossed determine the power that will be transmitted assuming that the maximum tension and belt speed are unchanged.

3 A ship is secured to its berth via a rope wound round a capstan. The ship is capable of pulling on the rope with a force of 20 kN, and the pull on the free end of the rope is 120 N. If the coefficient of friction between the rope and the capstan is 0·3, calculate the number of turns of the rope that are required to ensure that the ship is securely berthed.

4 A rope drive consists of four ropes which transmit power to a grooved pulley of diameter 1·2 m rotating at 300 rev/min. The groove angles are 40° and the angle of lap is 160°. Calculate the power that can be transmitted if the maximum tension in each rope is limited to 800 N. Assume $\mu = 0 \cdot 4$.

5 Two pulleys, centre distance 0·75 m, are connected by a flat belt with an initial tension of 20 N. The driving pulley is 250 mm diameter and the speed reduction is 4 to 1. For a belt speed of 4 m/s calculate the maximum power transmitted. $\mu = 0 \cdot 28$.

8.5 Centripetal tension

The analysis of belt drives used so far makes the assumptions that the mass of the belt is negligible, and that the belt speeds are low. If either of these conditions is not true then a further tension, in addition to T_1 and T_2, is experienced by the belt due to centripetal effects as it follows a circular path around the pulleys. Consider a small element of the belt of mass m per unit length which subtends an angle $\delta\theta$ at the centre of rotation, radius r. Let the belt speed be v m/s, and T_c the extra tension due to centripetal effects.

The length of the element $= r\,\delta\theta$
the mass of the element $= mr\,\delta\theta$

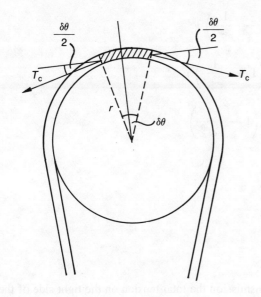

Fig. 8.5

the centripetal acceleration $= \dfrac{v^2}{r}$

and so the centripetal force $= mr\,\delta\theta\,\dfrac{v^2}{r}$

Referring to Fig. 8.5 shows that the centripetal force can only be supplied by the radial components of T_c.

Hence $2T_c \sin \dfrac{\delta\theta}{2} = mr\,\delta\theta\,\dfrac{v^2}{r}$

since $\delta\theta$ is very small, $\sin \dfrac{\delta\theta}{2} \simeq \dfrac{\delta\theta}{2}$

Whence $\underline{T_c = mv^2}$ \hfill (8.8)

The centripetal tension T_c is thus independent of the angle of lap and so is constant throughout the full length of the belt.

The total tensions on each side of the belt are thus:

$T_1 + T_c$ on the tight side, and
$T_2 + T_c$ on the slack side.

Note that since T_c is added to *both* sides then the difference in tensions is still $T_1 - T_2$, and the formula for power transmission (equation 8.5) is unchanged. Using equation (8.6) for the transmission of maximum power,

$$\text{max. power, } P = T_1\left(1 - \frac{1}{e^{\mu\theta}}\right)v$$

now, T_1 = total tension in belt on tight side minus centripetal tension or

$T_1 = T - T_c.$

$$\therefore P = (T - T_c)\left(1 - \frac{1}{e^{\mu\theta}}\right)v$$

$$= (T - mv^2)\left(1 - \frac{1}{e^{\mu\theta}}\right)v$$

$$= (Tv - mv^3)\left(1 - \frac{1}{e^{\mu\theta}}\right)$$

For a maximum $\dfrac{dP}{dv} = 0$

or $\qquad T - 3mv^2 = 0$
whence $\qquad\qquad T = 3mv^2$

but $T_c = mv^2$

$$\therefore T = 3 T_c \qquad\qquad\qquad\qquad (8.9)$$

Hence for maximum power transmission the total tension on the tight side of the belt equals three times the centripetal tension.

Example 8.3

A flat belt 80 mm wide and 5 mm thick is made from material having a density of 10^3 kg/m^3. The belt is used to transmit power to a pulley where the angle of lap is 140° and the coefficient of friction is 0·25. The maximum stress in the belt is not to exceed 2 MN/m^2. Determine the maximum power that the belt can transmit and the speed of the belt in this condition.

Maximum permissible belt tension = stress × area
$$= 2 \times 10^6 \times 0\cdot1 \times 0\cdot005$$
$$= \underline{1000 \text{ N}}$$

Mass of unit length of belt, $m = 10^3 \times 0\cdot1 \times 0\cdot005 \times 1$
$$= \underline{0\cdot5 \text{ kg/m}}$$

At maximum power

the centripetal tension, $T_c = \dfrac{\text{max. belt tension}}{3}$

$$= \frac{1000}{3}$$

$$= \underline{333\cdot3 \text{ N}}$$

From equation (8.8)

$$T_c = mv^2$$

$$\therefore v^2 = \frac{T_c}{m} = \frac{333\cdot3}{0\cdot5}$$

or $v = \underline{25 \cdot 8 \text{ m/s}}$

$$\theta = 140° = \frac{140 \cdot 2\pi}{360} \text{ radians}$$

$$= 2 \cdot 44 \text{ radians}$$

From equation (8.3)

$$\frac{T_1}{T_2} = e^{\mu\theta}$$

$$= e^{0 \cdot 25 \times 2 \cdot 44}$$

$$= 1 \cdot 84$$

and $T_1 = $ max. allowable tension $- T_c$

$$= 1000 - 333 \cdot 3$$

$$= 666 \cdot 7 \text{ N}$$

$$\therefore T_2 = \frac{T_1}{1 \cdot 84} = \frac{666 \cdot 7}{1 \cdot 84} = \underline{362 \cdot 3 \text{ N}}$$

$$\therefore \text{ maximum power} = (T_1 - T_2) \, v$$

$$= (666 \cdot 7 - 362 \cdot 3) \, 25 \cdot 8$$

$$= 7853 \cdot 5 \text{ watts}$$

$$= \underline{7 \cdot 85 \text{ kW}}$$

Guided solution

A V-belt transmits power from a pulley 900 mm diameter rotating at 300 rev/min to a pulley of 600 mm diameter. The pulley groove angles are 40° and the distance between the pulley centres is $1 \cdot 5$ m. The belt has a cross-sectional area of 300 mm^2 and a density of 10^3 kg/m^3. Calculate the maximum power that is transmitted if the maximum belt tension is $1 \cdot 6$ kN.

Step 1.   Using a diagram of the belt system and simple trigonometry evaluate the angle of lap on the smaller pulley (i.e. the one on which slip will occur first).

($2 \cdot 94$ rads)

Step 2.   Calculate the ratio of the effective tensions (i.e. those causing rotation) remembering that this is a V-belt.   ($T_1/T_2 = 13 \cdot 18$)

Step 3.   Using the larger pulley rotating at the given speed calculate the belt speed.

(14.14 m/s)

Step 4.   Calculate the mass per unit length of the belt.   ($0 \cdot 9$ kg/m)

Step 5.   Using the calculated belt speed, determine the centripetal tension, T.

(180 N)

Step 6.   Hence find the effective tension T_1 on the tight side of the belt.   (1420 N)

Step 7.   Calculate T_2 using the result of Step 2.   ($107 \cdot 74$ N)

Step 8.   Calculate the power transmitted.   ($18 \cdot 56$ kW)

Problems 8.2

1 Two pulleys of $0 \cdot 5$ m diameter and $0 \cdot 25$ m diameter have centres $1 \cdot 5$ m apart and are connected by an open belt. The larger pulley rotates at 320 rev/min and the maximum allowable tension in the belt is 900 N. If the belt has a mass of $0 \cdot 4$ kg per metre length and $\mu = 0 \cdot 22$ calculate the maximum power which may be transmitted.

2 Two pulleys of equal diameter are connected by a flat belt 100 mm wide and 6 mm thick, and of density 10^3 kg/m^3. The coefficient of friction between the belt and pulleys is $0 \cdot 3$ and the maximum permissible stress in the material of the belt is $1 \cdot 5$ MN/m^2. Calculate the maximum power that the system can transmit and the corresponding belt speed.

3 A flat belt is used to drive the pulley of an air compressor. The belt has a cross-section of 75 mm $\times$ 5 mm and a density of $1 \cdot 1 \times 10^3$ kg/m^3. The angle of lap of the belt on the pulley of the compressor is $164°$, and the maximum belt tension is 15 N per mm width. Calculate
   a the speed of the driving pulley in rev/min when maximum power is being transmitted
   b the value of the maximum power.
   Take $\mu = 0 \cdot 31$.

4 A rope drive is used to transmit power between two pulleys of diameter $2 \cdot 5$ m and $1 \cdot 6$ m and whose centres are 5 m apart. The groove angle of each pulley is $38°$ and the rope has a mass of $2 \cdot 6$ kg/m. Assuming that the coefficient of friction is $0 \cdot 27$, and that the maximum allowable tension in the rope is 20 kN, determine the maximum power that can be transmitted and the corresponding belt speed.

5 A belt-drive system consists of five V-belts connecting two pulleys on shafts $1 \cdot 5$ m apart. The driving pulley is 300 mm diameter and it rotates at 750 rev/min. The driven pulley rotates at $\frac{1}{4}$ of the speed of the driving pulley. Each belt has a mass of $0 \cdot 4$ kg/m and $\mu = 0 \cdot 3$. The groove angle is $40°$. If the maximum allowable tension in each belt is limited to 775 N, determine the power that can be transmitted.

6 A belt drive is required to transmit 250 kW with a belt speed of 25 m/s between pulleys of equal diameter. If a flat belt is used the ratio of $T_1/T_2 = 3 \cdot 5$. If a V-belt is used, the ratio of $T_1/T_2 = 5 \cdot 8$. Calculate the load exerted on the bearings of the pulleys in each case.

7 Double V-belts are assembled between two pulleys $0 \cdot 7$ m apart. The driving pulley has a diameter of 150 mm and rotates at 1400 rev/min. The driven pulley has a diameter of 200 mm and the pulley groove angles are $38°$. The mass per unit length of the material of each belt is $0 \cdot 3$ and the coefficient of friction between belts and pulleys is $0 \cdot 2$. Determine the maximum power that can be transmitted at the given speed.

8 A belt drive is required to transmit 30 kW at a belt speed of 12 m/s. The angle of lap on the smaller pulley is $170°$ and the coefficient of friction between the belts and pulley is $0 \cdot 36$. Calculate the number of belts required to transfer the given power if each belt is 100 mm wide by 5 mm thick, and the maximum allowable stress in each belt is 2 MN/m^2.

8.6   Friction clutches

A friction clutch is used to transmit power from one shaft to another co-axial shaft. The main advantage being that the clutch may be readily disengaged, when power is not required to be transmitted, and engaged again when power is required. Another advantage is that gradual engagement of the clutch is possible so allowing a smooth transfer of power and avoiding sudden impact torques.

Friction clutches may be of the flat plate type or the cone type. Flat plate clutches may have just one pair of friction surfaces, but more usually consist of a plate with friction material on each side. This latter arrangement constitutes two pairs of friction surfaces, and has the advantage that it is easily replaced when worn (Fig. 8.6).

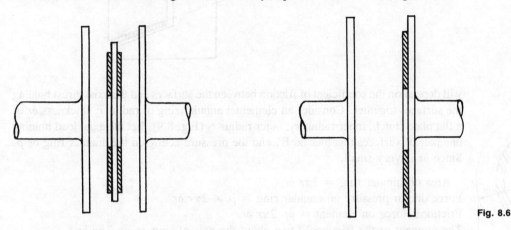

Fig. 8.6

Additional power can be accommodated without increasing the overall diameter of a flat plate clutch by increasing the number of pairs of friction surfaces. These multi-plate clutches are found in situations where space is at a premium but the power transmission required exceeds that of a single clutch plate (Fig. 8.7).

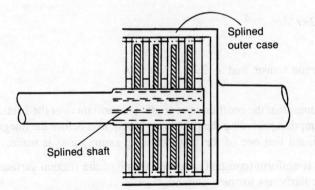

Splined outer case

Splined shaft

Fig. 8.7 A multi-plate clutch showing eight pairs of friction surfaces (shaded) on a splined shaft; intervening metal plates are splined to an outer case.

The cone clutch consists of a male member in the form of a frustum of a cone which usually carries the friction material, and a female member which is usually the driving component (Fig. 8.8).

In order to calculate the power that can be transmitted by a flat plate clutch it is necessary to first calculate the torque that can be conveyed without slip. This torque

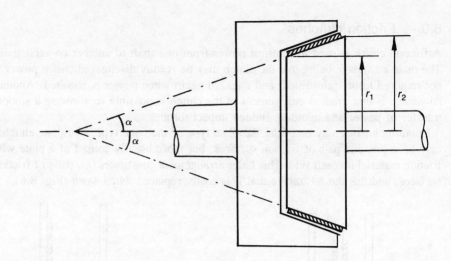

Fig. 8.8

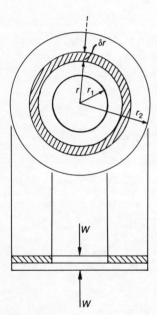

Fig. 8.9

will depend on the coefficient of friction between the surfaces and the axial thrust holding the surfaces together. Consider an elemental annular ring of radius r, thickness δr of a flat plate clutch, inner radius r_1, outer radius r_2 (Fig. 8.9). Let the axial load holding one pair of surfaces together be W, and the pressure acting on the annular ring be p. Since δr is very small,

Area of annular ring $= 2\pi r\, \delta r$
Force due to pressure on annular ring $= p \times 2\pi r\, \delta r$
Frictional force on element $= \mu p\, 2\pi r\, \delta r$
The moment of the frictional force about the axis of spin $= \mu p\, 2\pi r\, \delta r\; r$
$\qquad\qquad\qquad\qquad\qquad\qquad\qquad\qquad\qquad = \mu p\, 2\pi r^2\, \delta r$

The sum of the moments of all the frictional forces acting on all such elements between r_1 and r_2

$$= \int_{r_1}^{r_2} \mu p\, 2\pi r^2\, dr$$

$$\therefore \text{ total torque transmitted} = 2\mu\pi \int_{r_1}^{r_2} p r^2\, dr \qquad (8.10)$$

Equation (8.10) assumes that the coefficient of friction is constant over the friction surface. A further assumption regarding the pressure p is required before the integral can be evaluated. It is usual that one of the two following assumptions is made:

a That the pressure is uniform (constant) over the whole of the friction surface. This assumption is particularly apt for new clutches.
b that the rate of wear is constant over the whole of the friction surface. This assumption is more apt for worn clutches.

The rate of wear is proportional to the pressure and to the radius, and is thus usually assumed to be proportional to the product of the two.

Thus

rate of wear $\alpha\; pr = $ constant, k_r

Assuming uniform pressure, equation 8.10 becomes

$$\text{torque} = 2\mu p\pi \int_{r_1}^{r_2} r^2\, dr$$

$$= \tfrac{2}{3}\mu p\pi\, [r_2{}^3 - r_1{}^3] \tag{8.11}$$

Assuming uniform wear equation (8.10) becomes

$$\text{torque} = 2\mu pr\pi \int_{r_1}^{r_2} r\, dr$$

$$= 2\mu k_p\pi \left[\frac{r_2{}^2 - r_1{}^2}{2}\right] \tag{8.12}$$

The pressure p can be related to the axial load W.

(i) If the pressure is uniform over the friction surfaces then

$$p = \frac{\text{load}}{\text{cross-sectional area}} = \frac{W}{\pi(r_2{}^2 - r_1{}^2)}$$

Substituting this into equation (8.11) gives

$$\text{torque} = \tfrac{2}{3}\mu W \left[\frac{r_2{}^3 - r_1{}^3}{r_2{}^2 - r_1{}^2}\right] \tag{8.13}$$

for a flat plate assuming uniform pressure.

(ii) If the wear is uniform then $pr = $ constant, k_r.
The force on the annular element of Fig. 8.8 was shown to be given by

$$\text{force of element} = p \times 2\pi r\, \delta r$$

$$\text{total axial load } W = \int_{r_1}^{r_2} p\, 2\pi r\, dr$$

$$= \int_{r_1}^{r_2} 2\pi\, k_r\, dr$$

$$= 2\pi k_r\, [r_2 - r_1]$$

re-arranging

$$k_r = \frac{W}{2\pi\, (r_2 - r_1)}$$

Replacing k_r in equation (8.12) gives

$$\text{torque} = 2\mu\pi \; \frac{W}{2\pi \; (r_2 - r_1)} \left[\frac{r_2{}^2 - r_1{}^2}{2} \right]$$

$$\text{or torque} = \mu W \frac{(r_2 + r_1)}{2} \tag{8.14}$$

for a flat plate assuming uniform wear.

For a cone clutch with semi-vertex angle equal to α (Fig. 8.7) equations (8.13) and (8.14) are modified thus for uniform pressure:

$$\text{Torque} = \tfrac{2}{3} \frac{\mu W}{\sin \alpha} \left[\frac{r_2{}^3 - r_1{}^3}{r_2{}^2 - r_1{}^2} \right] \tag{8.15}$$

for uniform wear:

$$\text{torque} = \frac{\mu W}{\sin \alpha} \frac{(r_2 + r_1)}{2} \tag{8.16}$$

Using the assumption of uniform pressure will invariably give a higher value for the friction torque than that given by assuming uniform wear. Unless otherwise stated the uniform wear theory for a worn clutch is usually used so that a margin of safety in calculations is allowed.

Example 8.4

The clutch of a motor car consists of a single plate with friction surfaces on both sides and with inner diameter 250 mm and outer diameter 400 mm.

a Assuming a uniform pressure of 170 kN/m² acts over the whole of the friction surfaces and $\mu = 0\cdot28$, calculate the power transmitted at 450 rev/min.

b If this same power is required to be transmitted when the clutch becomes worn, and the assumption of uniform wear applies, ($\mu = 0\cdot24$) calculate the change in value of the axial load.

a The axial load W is transmitted across each pair of friction surfaces.

Hence $W = \text{pressure} \times \text{area}$

$$= 170 \times 10^3 \times \pi \frac{(0\cdot4^2 - 0\cdot25^2)}{4}$$

$$= \underline{13018 \text{ N}}$$

From equation (8.13) for uniform pressure:

$$\text{torque} = \tfrac{2}{3} \mu W \left[\frac{r_2{}^3 - r_1{}^3}{r_2{}^2 - r_1{}^2} \right]$$

$$= \tfrac{2}{3} \times 0 \cdot 28 \times 13018 \left[\frac{0 \cdot 2^3 - 0 \cdot 125^3}{0 \cdot 2^2 - 0 \cdot 125^2} \right]$$

$$= \underline{603 \text{ Nm}}$$

The power transmitted by *two* pairs of contact surfaces

$$= 2 \times \frac{2\pi NT}{60} \text{ watts}$$

$$= \frac{2 \times 2\pi \times 450 \times 603}{60} = \underline{56 \cdot 8 \text{ kW}}$$

b To transmit the same power the torque must be unchanged, hence from equation (8.14) for uniform wear:

$$\text{torque} = \mu W \frac{(r_2 + r_1)}{2}$$

or

$$603 = 0 \cdot 24 \, W \frac{(0 \cdot 2 + 0 \cdot 125)}{2}$$

whence $W = 15\,461$ N

The change in axial load is thus $15\,461 - 13\,018$ N

$$= \underline{2443 \text{ N}}$$

Example 8.5

A cone clutch is used to transmit 20 kN at 800 rev/min. The cone angle is 30° and the mean diameter of the friction surface is 250 mm. The intensity of pressure normal to the friction surfaces is 120 kN/m² and $\mu = 0 \cdot 3$. Calculate the axial load required and the width of the conical friction surface.

Assuming uniform wear, then from equation (8.16)

$$\text{torque} = \frac{\mu W}{\sin \alpha} \frac{(r_2 + r_1)}{2}$$

$\dfrac{r_2 + r_1}{2}$ is the mean radius $= 0 \cdot 125$ m

Thus $\quad \dfrac{20 \times 10^3 \times 60}{2\pi \times 800} = \dfrac{0 \cdot 3 \, W}{\sin 15°} \times 0 \cdot 125$

whence $W = \underline{1647 \cdot 7 \text{ N}}$

The axial load W is equal to the resolved part of the normal pressure in the axial direction multiplied by the area of friction surface.

i.e. $W = p \sin 15° \, (2\pi \times 0\cdot125 \times \text{width})$

i.e.   width of friction surface $= \dfrac{W}{p \sin 15 \times 2\pi \times 0\cdot125}$

$= \dfrac{1647\cdot7}{120 \times 10^3 \sin 15° \times 2\pi \times 0\cdot125}$

$= 0\cdot0675 \text{ m}$

$= \underline{67\cdot5 \text{ mm}}$

Example 8.6

An electric motor develops a constant torque of 50 Nm and is connected through a simple plate clutch with one pair of friction surfaces to a machine. The moment of inertia of the rotating parts of the motor and machine are 20 kgm² and 48 kgm² respectively. The clutch friction surfaces have an inner radius of $0\cdot25$ m and the outer radius is $0\cdot5$ m; $\mu = 0\cdot24$. The clutch surfaces are held in contact by a system of springs and the total axial spring force is $1\cdot5$ kN. The motor is started and run up to a speed of 500 rev/min when the clutch is engaged. Assuming uniform wear calculate

a the time taken for slipping of the clutch to cease
b the common speed of the motor and machine when slipping has just ceased
c the energy dissipated as heat during the engagement of the clutch

Assuming limiting friction while the clutch is slipping

$$\text{torque during slipping} = \mu W \frac{(r_2 + r_1)}{2}$$

$$= 0\cdot24 \times 1\cdot5 \times 10^3 \frac{(0\cdot75)}{2}$$

$$= \underline{135 \text{ Nm}}$$

This torque is applied equally to the machine and to the motor, and causes the machine to accelerate up to the common speed, and the motor to decelerate down to the common speed.

If ω rad/s is the common speed and t s the time that slipping takes place then applying Newton's second law for rotating parts

$$T = I\alpha$$
For the machine $T = 135$ Nm

$$\alpha = \frac{\omega}{t} \text{ rad/s}$$

and $I = 48 \text{ kgm}^2$

$$\therefore \quad 135 = 48 \frac{\omega}{t}$$

$$\text{or} \quad \omega = 2 \cdot 8125\, t \tag{i}$$

For the motor the net torque is the difference between 135 Nm and the motor torque

i.e. $\quad T = 135 - 50 = 85\ \text{Nm}$

$$= \frac{500 \times \dfrac{2\pi}{60} - \omega}{t} = \frac{16 \cdot 67\pi - \omega}{t}$$

$$I = 20\ \text{kgm}^2$$

hence $\quad 85 = 20 \dfrac{(16 \cdot 67\pi - \omega)}{t}$

Substituting for ω

$$85\,t = 20\,(16 \cdot 67\pi - 2 \cdot 8125\,t)$$
$$85\,t + 56 \cdot 25\,t = 1047 \cdot 1$$
$$\underline{t = 7 \cdot 4\ \text{s}}$$

from (i) the common speed $\omega = 2 \cdot 8125 \times 7 \cdot 4$
$$= \underline{20 \cdot 81\ \text{rad/s}}$$

Applying the law of conservation of energy to system:
initial energy of motor + work energy done by torque during slipping = final energy of motor and machine + heat energy dissipated.

Initial energy of motor $= \frac{1}{2} I\, \omega^2 = \frac{1}{2} \times 20 \times \left(\dfrac{500 \times 2\pi}{60} \right)^2$

$$= \underline{27 \cdot 4\ \text{kJ}}$$

Work energy done by torque $= T\theta$ where θ is the angle turned through and is given by

$$\theta = \frac{\omega_1 + \omega_2}{2} \cdot t$$

$$= \frac{\left(\dfrac{500 \times 2\pi}{60} + 20 \cdot 81 \right)}{2} \times 7 \cdot 4$$

$$= 270 \cdot 7\ \text{rads}$$

Thus work energy input by torque $= 50 \times 270 \cdot 7$
$$= \underline{13 \cdot 54\ \text{kJ}}$$

Final energy of motor + machine $= \frac{1}{2} I\, \omega^2$

$$= \tfrac{1}{2}(48 + 20)(20 \cdot 81)^2$$
$$= 14 \cdot 72 \text{ kJ}$$

Energy dissipated as heat $\qquad = 27 \cdot 4 + 13 \cdot 54 - 14 \cdot 72$
$$= \underline{26 \cdot 22 \text{ kJ}}$$

Guided solution

A multi-plate clutch has inner and outer diameter of 200 mm and 300 mm respectively, and is required to transmit 75 kW of power from an engine at 1000 rev/min. If a uniform pressure of 120 kN/m^2 is assumed and $\mu = 0 \cdot 22$, calculate the axial load necessary and the number of plates required.

Step 1. Calculate the torque that will transmit 75 kW at 1000 rev/min.

(716·2 Nm)

Step 2. Calculate the axial load from the known uniform pressure. (4712·4 N)

Step 3. Comparing the calculated torque from step 1 with the torque found by using the formula for uniform pressure, determine the number of friction surfaces required. Hence find the number of plates required assuming both sides of each plate are effective. (3)

Problems 8.3

1 A single plate clutch having both sides effective is used to transmit 10 kN from an engine rotating at 1000 rev/min. The internal and external diameters of the friction surfaces are 150 mm and 200 mm respectively and the coefficient of friction is 0·28. Assuming that the rate of wear is uniform calculate the required axial load.

2 A plate clutch consists of two pairs of friction surfaces which are held in contact by six springs each of stiffness 20 kN/m. The initial compression of each spring is 10 mm, and the inner and outer radii of the friction surfaces are 100 mm and 150 mm respectively. If $\mu = 0 \cdot 26$ calculate the power that can be transmitted at 1500 rev/min assuming uniform pressure.

3 After some time in operation the friction surfaces of the clutch in question 2 are found to have worn away by 1·5 mm. Assuming the wear was uniform and μ is unchanged calculate the change in power that can be transmitted at the same speed.

4 A multi-plate clutch is required to transmit 30 kW at 500 rev/min. The inner and outer diameters of the friction surfaces are 175 mm and 225 mm respectively and $\mu = 0 \cdot 29$. The maximum allowable pressure between the friction surfaces is 160 kN/m^2, and uniform wear is assumed. Calculate the axial load on the plates and the number of pairs of friction surfaces required.

5 An engine transmits 40 kW at 1400 rev/min through a cone clutch of mean diameter 250 mm and vertex angle 25°. Assuming uniform wear, $\mu = 0 \cdot 3$ and the intensity of pressure normal to the friction surface is 150 kN/m^2, determine the width of the friction surface and the axial load required.

6 A single plate clutch is used to connect an electric motor and a machine. The friction surfaces have an internal and external diameter of 200 mm and 300 mm respectively and the coefficient of friction is $0\cdot3$. The axial load is 1200 N. The moments of inertia of the motor and machine are 50 kgm^2 and 70 kgm^2 respectively.

　　The electric motor is running at 1200 rev/min and develops a constant torque of 200 Nm when the clutch is engaged. Determine:

a the time taken for the motor and machine to reach a common speed

b the magnitude of the common speed

c the heat energy dissipated due to clutch slip

8.7   Dry plain bearings

A selection of plain bearings is shown in Fig. 8.10. These types of bearings are used to locate rotating elements such as shafts and yet to affect the power transmission by the shaft as little as possible. In each case the torque required to overcome the dry friction of the bearing can be found by using the appropriate equations already derived for friction clutches, i.e. equations (8.13), (8.14), (8.15) and (8.16). In these instances the power developed is lost to the system rather than being transmitted. The assumptions of uniform pressure or uniform wear still apply but it is customary in the case

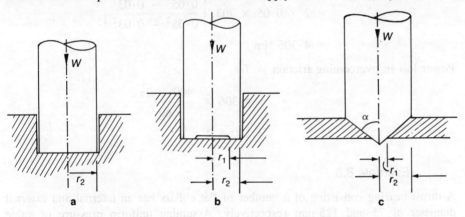

a

b

c

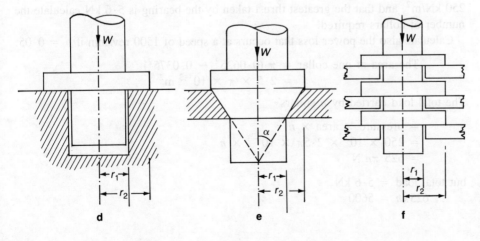

d

e

f

Fig. 8.10 a footstep bearing
b annular footstep **c** conical pivot
d collar bearing **e** conical collar
f multi-collar bearing

of bearings to assume uniform pressure, unless otherwise stated, as this will give the largest figure for power loss and so err on the side of safety.

Example 8.7

A flat collar bearing has internal and external diameter of 60 mm and 100 mm respectively and the coefficient of friction is $0 \cdot 05$. Assuming the pressure is uniform at 140 kN/m^2 calculate the power lost in overcoming friction in the bearing at a rotational speed of 400 rev/min.

$$\text{Axial load} = \text{pressure} \times \text{area of bearing surface}$$
$$= 140 \times 10^3 \times \pi \times (0 \cdot 05^2 - 0 \cdot 03^2)$$
$$= \underline{703 \text{ N}}$$

From equation (8.13)

$$\text{friction torque} = \tfrac{2}{3} \mu W \left[\frac{r_2^3 - r_1^3}{r_2^2 - r_1^2} \right]$$

$$= 2 \times 0 \cdot 05 \times 703. \left[\frac{0 \cdot 05^3 - 0 \cdot 03^3}{0 \cdot 05^2 - 0 \cdot 03^2} \right]$$

$$= 4 \cdot 306 \text{ Nm}$$

Power lost in overcoming friction $= T\omega$

$$= 4 \cdot 306 \times \frac{400 \times 2\pi}{60}$$

$$= 180 \cdot 4 \ W$$

Example 8.8

A thrust bearing consisting of a number of flat collars has an internal and external diameter of 75 and 125 mm respectively. Assuming uniform pressure of value 250 kN/m^2, and that the greatest thrust taken by the bearing is $5 \cdot 6$ kN calculate the number of collars required.

Calculate also the power loss that occurs at a speed of 1500 rev/min if $\mu = 0 \cdot 05$.

$$\text{The area of one collar} = \pi \, (0 \cdot 0625^2 - 0 \cdot 0375^2)$$
$$= 2 \cdot 5 \times \pi \times 10^{-3} \text{ m}^2$$

The total load carried by n collars

$$= \text{pressure} \times \text{area} \times n$$
$$= 250 \times 10^3 \times 2 \cdot 5\pi \times 10^{-3} \times n$$
$$= 625 \ \pi n \text{ N}$$

but total load $= 5 \cdot 6$ kN
$$\therefore \ 625\pi n = 5600$$

$$n = \frac{5600}{625\pi} = 2 \cdot 85$$

Hence 3 collars are required.

The friction torque, $T = \frac{2}{3} \mu W \left[\dfrac{r_2{}^3 - r_1{}^3}{r_2{}^2 - r_1{}^2} \right]$

$$= \frac{2}{3} \times 0 \cdot 05 \times 5600 \times \left[\frac{0 \cdot 0625^3 - 0 \cdot 0375^3}{0 \cdot 0625^2 - 0 \cdot 0375^2} \right]$$

$$= \underline{14 \cdot 28 \text{ Nm}}$$

Power loss $= T\omega$

$$= \frac{14 \cdot 28 \times 2\pi \times 1500}{60}$$

$$= \underline{2 \cdot 24 \text{ kW}}$$

Problems 8.4

1 A vertical shaft 50 mm diameter runs in a simple footstep bearing. If the coefficient of friction between the shaft and the bearing housing is $0 \cdot 04$, and the pressure is assumed uniform at 100 kN/m^2, calculate the power loss at a shaft speed of 800 rev/min.

2 The axial thrust of 800 N in a vertical shaft rotating at 1000 rev/min is taken by a conical pivot of inner diameter 50 mm and outer diameter 80 mm and vertex angle 90°. The coefficient of friction between the bearing surfaces is $0 \cdot 02$. Calculate the power lost due to friction assuming uniform pressure.

3 A multi-collar bearing consists of two pairs of bearing surfaces of inner and outer diameter 50 mm and 75 mm respectively. Assuming that the intensity of pressure is uniform at 150 kN/m^2 and that $\mu = 0 \cdot 025$ calculate the power loss that occurs at 900 rev/min.

4 Determine the number of pairs of bearing surfaces required in a multi-collar bearing of inner and outer diameter 70 mm and 120 mm assuming that the intensity of pressure between pairs of surfaces is limited to 200 kN/m^2 and the maximum axial load is $8 \cdot 5$ kN. If $\mu = 0 \cdot 06$ calculate the power loss at 2000 rev/min.

8.8   Lubricated plain bearings

The friction that exists between the contact surfaces of dry plain bearings may be relieved by introducing a lubricant between the surfaces. In a fully lubricated bearing a thin but definable film of lubricant prevents contact of the bearing material so reducing the torque necessary to turn the shaft forming part of the bearing, and so reducing the power loss. In fact in a properly lubricated bearing the only torque required is that needed to overcome the natural resistance of the lubricant to move. The measure of how easily a fluid flows is its *viscosity*. The higher the viscosity of a fluid the more reluctant it is to flow and the greater the force required to overcome the resistance.

Consider, for example, a surface A moving with a velocity v over a stationary surface B with a small intervening gap of width x. (Fig. 8.11).

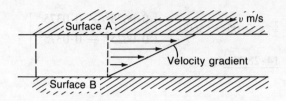

Fig. 8.11

The gap is filled with a lubricant which has zero velocity at surface B, but has a velocity of v m/s relative to B at surface A. Simple theory suggests that the layers of lubricant between B and A will move relative to each other in a shearing action and will have velocities which increase linearly with distance across the film of lubricant. That is the velocity gradient across the lubricant is constant.

$$\text{Thus in this case;} \quad \frac{dv}{dx} = \frac{\text{relative velocity}}{\text{thickness of lubricant film}}$$

$$\text{or, rate of shear} = \frac{v}{x}$$

Further since each layer of fluid is moving relative to its neighbour a shear stress τ exists between the layers which opposes this tendency to move. The ratio of this shear stress to the velocity gradient is termed the **coefficient of dynamic viscosity**, η.

Thus:
$$\eta = \frac{\tau}{dv/dx} \quad \frac{N_s}{m^2} \tag{8.17}$$

$$\text{or} \quad \tau = \eta \frac{v}{x} \tag{8.18}$$

where v is the relative velocity of bearing surfaces, and x is the thickness of film of lubricant.

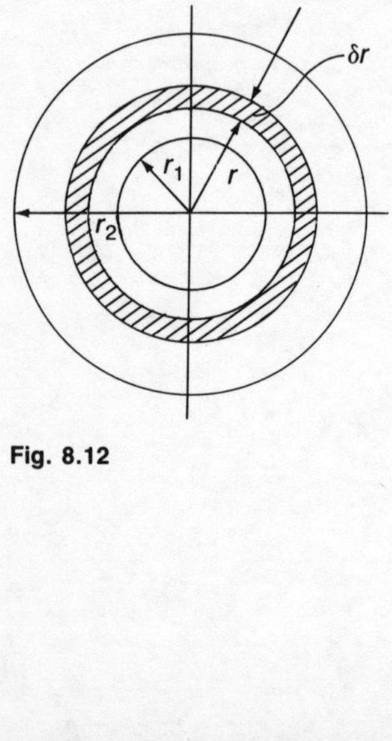

Equation (8.18) is often quoted as Newton's Law of Viscosity which is simply stated; shear stress, τ, is directly proportional to the rate of shear. Fluids which obey this law are termed 'Newtonian Fluids' and these include most of the mineral-oil-based lubricants. The power required to overcome the viscous resistance of the lubricant is simply calculated. For example, the annular collar bearing shown (Fig. 8.12) has an internal radius r_1, and external radius r_2. Let the bearing be rotating at ω rad/s with a film of oil of thickness h metres and viscosity η between it and its housing.

Considering an annular elemental ring of radius r and thickness δr

Fig. 8.12

$$\text{relative velocity at radius } r, \quad v = \omega r$$

velocity gradient at radius $r = \dfrac{v}{h} = \dfrac{\omega r}{h}$

From equation (8.18)

viscous stress on element $= \eta \, \dfrac{\omega r}{h}$

$\therefore$ viscous force = viscous stress $\times$ area $= \eta \, \dfrac{\omega r}{h} \times 2\pi r \delta r$

Torque required to overcome viscous resistance

= viscous force $\times$ radius

$= \eta \, \dfrac{\omega r}{h} \times 2\pi r \, \delta r \, r$

$= \dfrac{2\pi \eta \, \omega r^3 \, \delta r}{h}$

Thus total torque required $= \dfrac{2\pi \eta \omega}{h} \displaystyle\int_{r_1}^{r_2} r^3 \, dr$

Power required $= T\omega = \dfrac{\pi \eta \omega^2}{2h} (r_2{}^4 - r_1{}^4)$ \hfill (8.19)

Example 8.9

A thrust bearing in the form of a flat collar of inner and outer diameter 75 mm and 120 mm respectively runs in its housing on a film of oil $0\cdot3$ mm thick.

The dynamic viscosity of the oil is $0\cdot1$ Ns/m^2. Calculate the power loss due to the viscous effects of the oil if the shaft integral with the bearing is rotating at 1500 rev/min.

Applying equation (8.19)

power loss $= \dfrac{\pi \eta \omega^2}{2h} (r_2{}^4 - r_1{}^4)$

$= \dfrac{\pi \times 0\cdot1}{2 \times 0\cdot0003} \left(\dfrac{1500 \times 2\pi}{60}\right)^2 (0\cdot06^4 - 0\cdot0375^4)$

$= 523\cdot6 \times 24674 \times 10\cdot9825 \times 10^{-6}$

$= \underline{141\cdot9 \text{ watts}}$

Example 8.10

A plain journal bearing has a diameter of 50 mm and a length of 25 mm. The radial clearance between the shaft and the bearing shell is $0\cdot5$ mm and this clearance is filled with oil having a dynamic viscosity of $0\cdot15$ Ns/m^2. Assuming that the shaft rotates

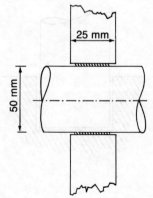

concentrically with the bearing calculate the power loss due to viscous drag at 2000 rev/min.

Relative velocity of shaft with respect to bearing $= r\omega$

$$= 0 \cdot 025 \times \frac{2000 \times 2\pi}{60}$$

$$= 5 \cdot 236 \text{ m/s.}$$

$$\text{Velocity gradient} = \frac{\text{velocity}}{\text{thickness of oil film}}$$

$$= \frac{5 \cdot 236}{0 \cdot 0005} = 10\ 472 \text{ /s}$$

Viscous stress $= \eta \times$ velocity gradient
$$= 0 \cdot 15 \times 10\ 472$$
$$= 1570 \cdot 8 \text{ N/m}^2$$

Viscous force $=$ viscous stress $\times$ area
$$= 1570 \cdot 8 \times \pi \times 0 \cdot 050 \times 0 \cdot 025$$
$$= 6 \cdot 17 \text{ N}$$

Torque resisting motion $=$ viscous force $\times$ radius
$$= 6 \cdot 17 \times 0 \cdot 025$$
$$= 0 \cdot 154 \text{ Nm}$$

Power loss due to this torque $= T\omega$

$$= 0 \cdot 54 \times \frac{2000 \times 2\pi}{60}$$

$$= \underline{32 \cdot 3 \text{ watts}}$$

Problems 8.5

1 A vertical shaft is supported by a simple footstep bearing of diameter 75 mm. The bearing is separated from the base of the housing by a film of oil of thickness 0·2 mm and viscosity 0·2 Ns/m². Ignoring the effects of the sides calculate the power absorbed by friction at a shaft speed of 1000 rev/min.

2 A journal bearing of diameter 70 mm and length 50 mm has a radial clearance of 0·08 mm with its housing. The clearance is filled with oil having a coefficient of viscosity 0·1 Ns/m². Calculate the power loss in the bearing when running concentrically at 2000 rev/min.

3 A multi-collar thrust bearing consists of six pairs of bearing surfaces of inner and outer diameter 60 mm and 100 mm respectively. Each pair of bearing surfaces is separated from its housing by a film of oil of thickness 0·3 mm and viscosity 0·1 Ns/m². Calculate the power loss in the bearing when rotating at 600 rev/min.

4 Find the force required to push a shaft of diameter 50 mm concentrically through a tube of diameter 50·5 mm and length 100 mm at a velocity of 2 m/s relative to the tube, if the annular space between the shaft and the tube is filled with oil having a viscosity of 0·3 Ns/m^2.

9 The dynamics of rigid body mechanisms

The contents of this chapter will enable you to construct velocity diagrams for standard link mechanisms, and to determine the relative velocity of individual links in the mechanism. You will also be able to determine the nature and purpose of machine flywheels, and the static and dynamic balancing of rotating masses.

9.1 Mechanisms

A mechanism is an assembly of components or links which are able to move relative to each other. A mechanism may be used to translate a movement or displacement of a point to another point some distance away, the movement being magnified, kept the same, or reduced as required. A mechanism can be used to translate a force or a torque from one point to another, i.e. to do work, and in this case the mechanism is called a **machine**.

The links of a mechanism may take very different forms. Examples of common links are levers, shafts, pulleys, cranks, pistons, sliders, connecting rods and bearings. Each link is capable of moving relative to its neighbouring links in a pre-determined manner dictated by the lengths, angles, points of location and other external constraints of all the links in the mechanism. In other words the *way* in which a mechanism moves is totally predictable.

A mechanism is formed by assembling links in series to form a **kinematic chain**. Most simple mechanisms can be reduced to one of two basic kinematic chains or their variations. These are the four-bar chain (or quadric cycle), and the slider-crank chain.

9.2 The four-bar chain

The four-bar chain consists of four rigid members joined together by frictionless pivots. The type of movement obtained from the chain depends on the length of the members and which of the members is chosen as a reference or datum. Usually the reference member is part of the frame of the mechanism or otherwise connected to 'earth'. In any case this member is assumed to be held motionless relative to the other members. For example consider the mechanisms of Fig. 9.1.

a   This is a 4-bar chain with the members of such length, so as to form a double lever mechanism, and the member AD the datum, i.e. the members AB and CD can only oscillate backwards and forwards.

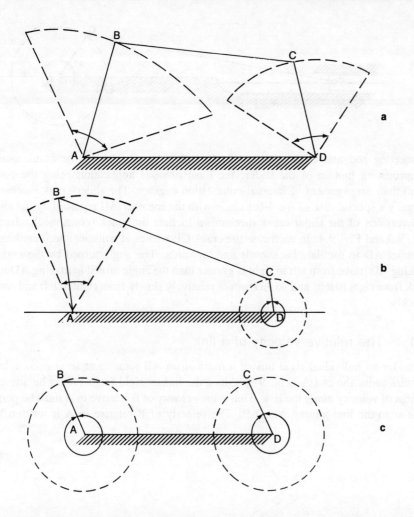

Fig. 9.1 Four-bar chains

b A 4-bar chain with members of such length so as to form a lever-crank mechanism. The member AB oscillates to and fro, while the member CD makes complete revolutions.

c A-4 bar chain forming a double-crank mechanism. The members AB and CD are of equal length and make complete revolutions.

The same 4-bar chain can be made to perform all of the above variations (providing that the lengths are carefully chosen) by making each member in turn the datum. This process of changing the characteristics of a 4-bar chain is known as **inversion**. Thus the 4-bar chain in Fig. 9.1a is a double-lever mechanism, but its inversion obtained by making BC the datum instead of AD is a lever-crank mechanism. The number of inversions possible is equal to the number of links in the chain.

9.3   The slider-crank mechanism

This kinematic chain consists of a crank, a connecting rod and a slider (Fig. 9.2). The member AD forms the datum; the slider B moves relative to the datum A; BC is the

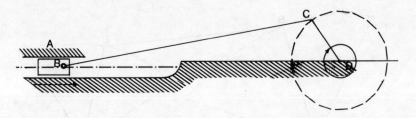

Fig. 9.2

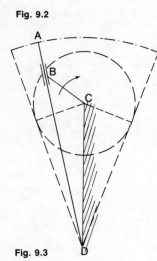

Fig. 9.3

connecting rod and CD is the crank. Complete revolutions of the crank cause a reciprocating motion of the slider, the most obvious application being the piston-crankshaft arrangement of internal-combustion engines. The slider-crank mechanism is really a special case of the 4-bar chain with the member AB replaced by the slider.

Inversions of the slider-crank mechanism include the quick-return mechanisms of Fig. 9.3 and Fig. 9.4. In each case the crank CB rotates at constant speed causing the member AD to oscillate backwards and forwards. The angle turned by the crank in making AD move from left to right is greater than the angle turned in making AD move back from right to left, and so AD moves relatively slowly from right to left and returns quickly.

9.4   The relative velocity of a link

Consider an individual rigid link of a mechanism AB rotating at an angular velocity ω rad/s about the end A (Fig. 9.5). Since the link is rigid there cannot be any component of velocity *along* the link. That is the velocity of B relative to A must be perpendicular to the line joining A and B. The velocity of B relative to A is written V_{BA},

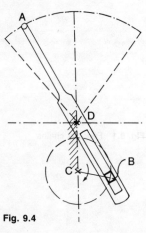

Fig. 9.4

Fig. 9.5

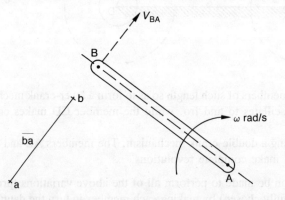

and similarly the velocity of A relative to B is written V_{AB}. The velocity V_{BA} can be represented by a vector b a which is a line drawn representing the magnitude and direction of the relative velocity. If the relative velocities of all the links in the mechanism are represented by their vectors then we have a **vector diagram** from which certain unknown quantities can be determined. This vector diagram is drawn for a given position of the mechanism and represents the velocities of the mechanism only at that moment in time. Thus it is usual in problems to be given an angle that one of the links makes with some reference line, so that using the dimensions of all the links a **space diagram** can be drawn to some convenient scale.

Example 9.1

In the 4-bar chain shown, the crank AB rotates at a constant angular velocity of 10 rad/s in a clockwise direction. AB = 0·2 m, BC = 0·4 m, CD = 0·3 m, AD = 0·55 m and angle DAB = 45°. For the given position draw the velocity vector diagram and determine the velocity of point C and the angular velocity of link BC.

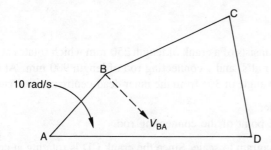

Enough information is supplied in the question to enable a space diagram to be accurately drawn to a suitable scale. The datum line AD is drawn first, say to a scale of $\frac{1}{4}$ full size, i.e. AD is drawn 137·5 mm long, followed by AB, 50 mm long at 45° to AD. From the established positions of B and D arcs of radii 100 mm and 75 mm are described respectively to establish the position of C. The velocity diagram is then drawn as described below.

The velocity of B relative to A is perpendicular to AB and has a magnitude given by:

$$V_{BA} = \omega_{BA} \times AB$$
$$= 10 \times 0·2$$
$$= 2 \text{ m/s.}$$

The vector representing this velocity, $\overline{ba}$, can now be drawn to a suitable scale, say 20 mm = 1 m/s, see the second diagram. Note that having established a position for a, b must be below and to the right of a. Note also the use of the lower-case letters on velocity diagrams.

The next link to consider is BC, all we know of its relative velocity is that it is perpendicular to BC, i.e. we know the direction only of the velocity vector. So starting from the point b on the vector diagram draw cb in a direction perpendicular to BC, see the third diagram. As yet the position of C cannot be positively established as the magnitude of cb cannot be calculated.

The link CD is pivoted at D, and as before, the relative velocity of C with respect to D is perpendicular to CD. However point D and point A are both fixed, i.e. their relative velocities are zero. This means that a and d occupy the same point on the velocity diagram. Starting at the established position for d then the vector cd can be drawn in direction only, see the fourth diagram. The position of c is then established.

The completed vector diagram can then be used to read off the scaled quantities required.

Thus the velocity of C relative to the fixed points is given by cd = 0·75 m/s at 8° to AD.

The angular velocity of BC = $\dfrac{V_{BC}}{BC} = \dfrac{bc}{BC}$

$$= \frac{1 \cdot 66}{0 \cdot 4} = 4 \cdot 15 \text{ rad/s}$$

From the point B, C appears to be moving to the left and so the rotation of BC is anti-clockwise.

Example 9.2

A slider-crank mechanism consists of a crank of length 250 mm which rotates at a constant angular velocity of 20 rad/s, and a connecting rod of length 900 mm. At the instant that the crank makes an angle of 30° from the inner dead-centre position calculate

a the velocity of the slider
b the velocity of the mid-point of the connecting rod.

Firstly we draw the space diagram to scale. Since the crank CD is rotating at constant angular speed the velocity of C with respect to D can be calculated.

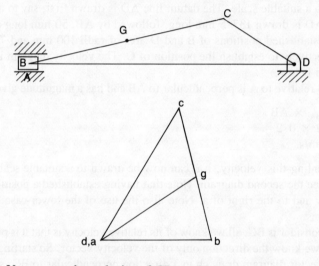

Thus V_{CD} = angular velocity of CD $\times$ CD
$= 20 \times 0 \cdot 25$
$= 5$ m/s.

Thus the magnitude and direction of V_{CD} allow the vector $\overline{cd}$ to be drawn completely. Taking the direction from the space diagram and using a suitable scale (say 20 mm $\equiv$ 1 m/s) to give a convenient length, draw $\overline{cd}$.

We have now fixed the point c and using this point as a start we can draw the *direction* of vector $\overline{bc}$ which is perpendicular to the link BC. As yet we do not know the length of this vector.

Finally the piston B is constrained to move horizontally relative to A, a fixed point. All fixed points occupy the same position on the velocity diagram, and so a is at d. The point b is then on a horizontal line through d, and the point of intersection of this line and the vector $\overline{bc}$, fixes b positively. From the resulting velocity diagram the velocity of the piston B (relative to a fixed point) is represented by $\overline{ba} = 2 \cdot 8$ m/s.

The mid-point of BC, G, appears as g at the mid-point of $\overline{bc}$ on the velocity diagram. The velocity of g (relative to a fixed point) is then represented by dg = 3·4 m/s.

Example 9.3

The quick-return mechanism shown consists of a crank CB which rotates clockwise at 500 rev/min and is connected by a slider B to a lever AD pivoted at D. A further lever connects A to a slider E which is caused to move slowly to the right and quickly to the left. AD = 0·5 m, BC = 0·1 m, AE = 0·35 m. At the instant that angle BCD = 45° calculate

a the velocity of the slider E
b the angular velocity of AE.

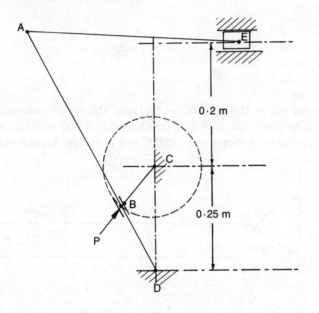

The angular velocity of BC = $500 \times \dfrac{2\pi}{60}$ rad/s

$$= 52 \cdot 36 \text{ rad/s.}$$

The velocity of B relative to C, $V_{BC} = \omega$. BC

$$= 52 \cdot 36 \times 0 \cdot 1$$
$$= 5 \cdot 24 \text{ m/s.}$$

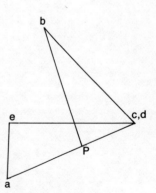

To a scale of 20 mm $\equiv$ 1 m/s draw the vector $\overline{bc}$ to represent V_{BC} in a direction perpendicular to BC. The points b and c on the velocity diagram are thus positively located.

Since C is a fixed point then the other fixed point d may also be located at the same place.

The lever AD pivots about D, and so V_{AD} may be represented by a vector $\overline{ad}$ whose direction is perpendicular to AD. The length of $\overline{ad}$ is not known, but a is to the left of d. To locate a, consider the slider B. At the instant under consideration let the slider be co-incident with a point P which is fixed on the lever AD. Point P too, pivots about D. The vector $\overline{pd}$ is on the same line as $\overline{ad}$.

Furthermore the slider B is moving relative to P in the direction DA at the instant under consideration. In fact the slider is constrained always to move along AD. The intersection of the vectors $\overline{pd}$ and $\overline{bp}$ will locate the point p. The point a is located on the line $\overline{dp}$ produced so that the ratio dp : pa is the same as the ratio DP : PA on the space diagram, because both points are on the same rigid line.

Finally the relative velocity of A with respect to E is perpendicular to AE and so the direction of $\overline{ae}$ is known. Since E is constrained to move horizontally in relation to a fixed point $\overline{ed}$ is a horizontal line and so point e is located.

From the finished velocity diagram, the velocity of the slide is given by

$$\overline{ed} = \underline{4\cdot7 \text{ m/s}}$$

The angular velocity of AE $= \dfrac{V_{AE}}{AE} = \dfrac{2}{0\cdot35} = \underline{5\cdot7 \text{ rad/s}}$

Guided solution

In the mechanism shown AB = 160 mm, BC = 540 mm, DE = 480 mm, and E is the mid-point of BC. The crank AB is rotating anticlockwise at 300 rev/min. At the instant shown, calculate the velocities of the slides C and D and the angular velocity of ED.

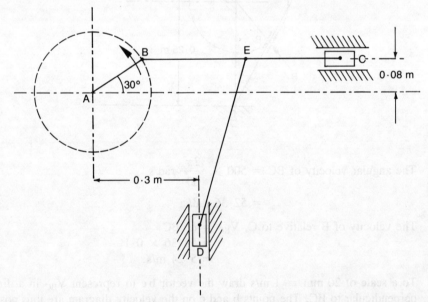

Step 1.   Using a suitable scale (say $\frac{1}{5}$ full size) draw an accurate space diagram of the mechanism. Leave space on the same page for the velocity diagram as this allows the directions of the vectors to be easily transferred using a ruler and set-square.

Step 2.   Calculate the angular velocity of AB in rad/s (31·42 rad/s), and use this value to find V_{BA} (5·03 m/s).

Step 3.   Using a suitable scale (say 10 mm ≡ 0·5 m/s) draw the vector $\overline{ba}$ 100·6 mm long in a direction perpendicular to AB on the space diagram using a set-square and ruler. Label the vector $\overline{ab}$ with a at the bottom.

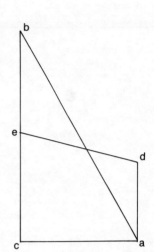

Step 4.   The slider C is constrained to move horizontally to the left relative to a fixed point such as A. Thus draw the direction of the vector $\overline{ac}$ horizontally through a and to the left of a.

Step 5.   The velocity of B relative to C is perpendicular to BC, that is vertical. Draw a vertical line through b to meet the horizontal drawn through a at c.

Step 6.   Locate e the mid-point of bc.

Step 7.   The slider D is constrained to move vertically upwards relative to a fixed point, so draw a vertical line through a. d lies on this line and above a.

Step 8.   Draw the vector $\overline{ed}$ in a direction perpendicular to ED to meet the vertical through a and d. The velocity diagram is now complete and should look as the velocity diagram here.

Step 9.   Determine the velocities of the sliders C and D by using the appropriate vectors from the diagram.

(Velocity of C = 2·68 m/s
Velocity of D = 1·75 m/s)

Step 10.   Determine V_{ED} from the diagram and use it to calculate the angular velocity of ED (5·73 rad/s).

Problems 9.1

1   In a simple slider-crank mechanism the crank is 100 mm long and the connecting rod is 300 mm long. The slider is on the same horizontal line as the centre of rotation of the crank. If the crank is rotating clockwise at 400 rev/min find:
   a   the velocity of the slider
   b   the angular velocity of the connecting rod when the crank is 30° from its outer dead-centre position.

2   In the four-bar chain shown the crank AB rotates at 500 rev/min clockwise. AB = 120 mm, BC = 360 mm, CD = 240 mm and E is the mid-point of BC. For the position shown calculate
   a   the velocity of the point E
   b   the angular velocity of the link BC.

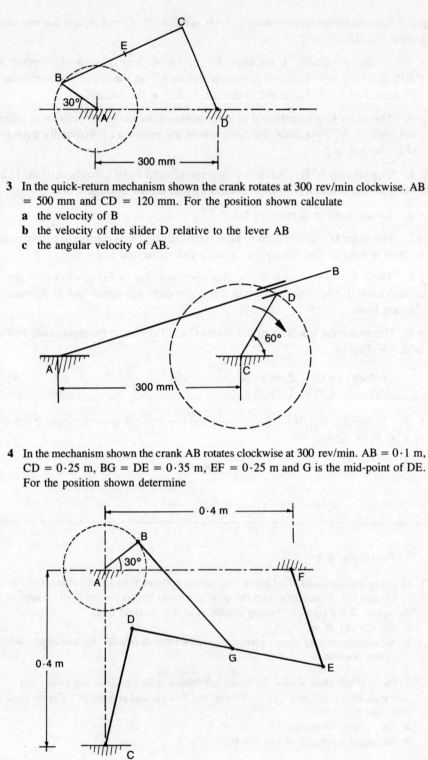

3 In the quick-return mechanism shown the crank rotates at 300 rev/min clockwise. AB
= 500 mm and CD = 120 mm. For the position shown calculate
   a the velocity of B
   b the velocity of the slider D relative to the lever AB
   c the angular velocity of AB.

4 In the mechanism shown the crank AB rotates clockwise at 300 rev/min. AB = 0·1 m,
CD = 0·25 m, BG = DE = 0·35 m, EF = 0·25 m and G is the mid-point of DE.
For the position shown determine

a the velocity of the point G
b the angular velocity of the link DE

5 The mechanism shown consists of a crank AB which rotates clockwise at 250 rev/min, connected by a link BC to a solid triangular link CDE pivoted at D. The link EF connects the triangular link to a slider F which is constrained to move in vertical guides. For the position shown, determine
a the velocity of the slider F
b the angular velocity of the triangular link CDE,
given that AB = 100 mm, BC = 270 mm, CD = 230 mm, DE = 140 mm, EF = 320 mm.

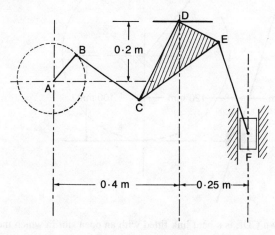

6 In the quick-return mechanism shown the crank AB rotates at 200 rev/min clockwise. AB = 200 mm, CD = 150 mm and DE = 500 mm. For the given position determine
a the velocity of the slider block E
b the angular velocity of the link DF.

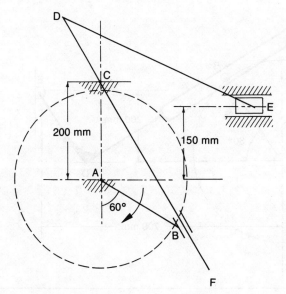

7 In the four-bar chain shown the link HG is pivoted at E, the mid-point of BC, and passes through a stationary pivoting slider F. The crank AB rotates clockwise at 120 rev/min and DC = 150 mm, BC = 260 mm, AB = 120 mm. For the position shown, determine
   a the velocity of the point E
   b the angular velocity of the link HG.

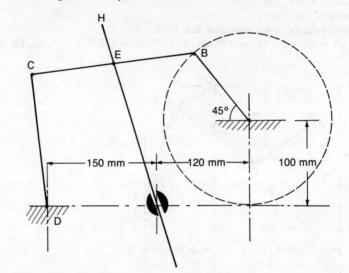

8 In the mechanism shown CDE is a bent link fitted with an open slot in which the end B of the crank AB is allowed to slide.
   The crank rotates at 180 rev/min clockwise and AB = 200 mm, CD = 800 mm, ED = 220 mm and angle CDE = 90°. For the position shown determine
   a the angular velocity of the link CDE
   b the velocity of the point D.

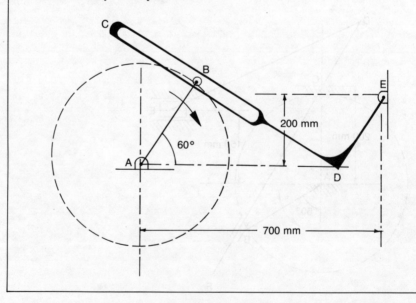

9.5   The function of flywheels

As previously mentioned the slider-crank mechanism forms the basis of the reciprocating engine. In these engines high pressure developed in the cylinder, caused by rapid burning of an air-fuel mixture or an injection of steam, produces a force on the piston F. This force together with a side force R, due to the reaction of the cylinder wall on the piston, produces a force in the connecting-rod P which in turn produces a turning moment T on the crankshaft, Fig. 9.6.

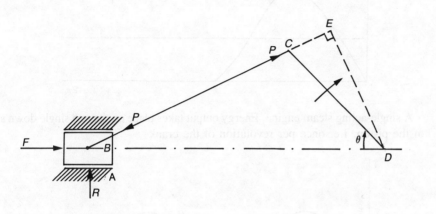

Fig. 9.6

The value of the turning moment produced is given by the product of P and the perpendicular distance of its line of action from the centre of crank rotation ED.

Thus, crankshaft torque $= P \times ED$.

Now the force in the connecting-rod is a variable quantity depending on the pressure and hence the force being developed in the cylinder. This force reaches a maximum value when the piston is at the top of its stroke, and a minimum when the piston is at the bottom of its stroke. Furthermore the length ED clearly depends on the position of the crank, being zero when the piston is at the top of its stroke (when $\theta = 0$), and a maximum when the angle BCD is a right-angle. Thus the turning moment or torque applied to the crank is a variable quantity which can be plotted against the crank angle θ to form a **turning-moment diagram**. Fig. 9.7 shows a typical turning-moment diagram.

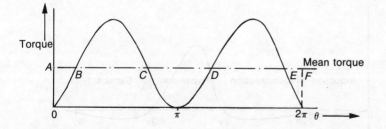

Fig. 9.7

The work done by the crankshaft torque per cycle, and hence the engine work done per cycle, is equal to the area under the turning-moment diagram. The mean crankshaft torque is represented by a horizontal straight line, AF, and in general this line will also represent the external torque which the engine is required to overcome. In practice it is more than likely that the external torque itself may be varying throughout the cycle. Other examples of turning-moment diagrams are shown in Fig. 9.8.

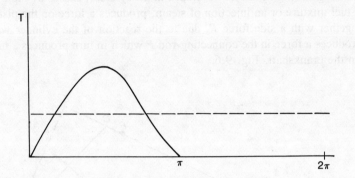

A single-acting steam engine. Energy output takes place on the one single down stroke of the piston, i.e. once per revolution of the crank.

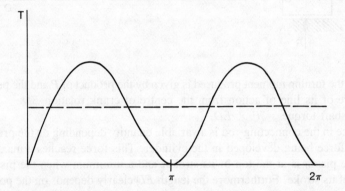

A two-stroke engine or double-acting steam engine. Energy output is achieved on each stroke of the piston.

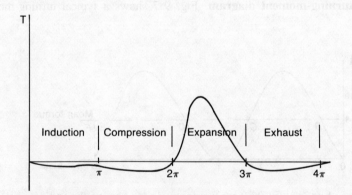

A single-cylinder four-stroke engine. Energy output is only achieved once in four strokes or two revolutions.

Fig. 9.8

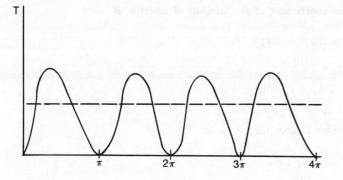

A four-cylinder four-stroke engine. Each cyclinder contributes an energy output for each half-revolution of the crank.

Under normal circumstances the work done, and hence the torque produced, by the engine just equals the work requirements of the load. If the torque produced by the engine is greater than that required by the external load, as it is for the portions B to C and D to E of the turning moment diagram in Fig. 9.7, then the engine will accelerate. If the torque produced by the engine is less than that required by the load, as it is for the portions A to B, C to D and E to F, then the engine will slow down.

To avoid these fluctuations in speed and energy output a flywheel is fitted to the crankshaft of all reciprocating type engines and mechanisms. The function of a flywheel is to absorb the excess energy produced by an engine, to store it and to release it when the energy requirements of the load exceed that produced by the engine. In other words a flywheel acts as a reservoir of energy which stores excesses and makes up deficiences as required. The result is a more even output of speed and power, although the exact extent of the smoothing out process depends on the size and weight of the flywheel used.

Suppose that ω_1 is the maximum speed attained by a flywheel during periods of absorption of energy, and ω_2 the minimum speed attained during periods of release of energy. If I is the moment of inertia of the flywheel then the greatest fluctuations of energy (E) that exists during a cycle is given by:

$$E = \tfrac{1}{2} I \omega_1^2 - \tfrac{1}{2} I \omega_2^2 \qquad\qquad (9.1)$$

Furthermore, if ω is the mean speed then the ratio

$$\frac{\text{maximum fluctuation of speed per cycle}}{\text{mean speed}}$$

is defined as the **coefficient of fluctuation of speed**, A.

So that

$$A = \frac{\omega_1 - \omega_2}{\omega} \qquad\qquad (9.2)$$

Also the ratio:

$$\frac{\text{maximum fluctuation of energy per cycle}}{\text{work done per cycle}}$$

is defined as the **coefficient of fluctuation of energy**, B.

So that $B = \dfrac{\frac{1}{2} I(\omega_1^2 - \omega_2^2)}{W}$ (9.3)

$\qquad = \dfrac{\frac{1}{2} I(\omega_1 + \omega_2)(\omega_1 - \omega_2)}{W}$

for small variations in speed $\omega_1 + \omega_2 \simeq 2\omega$

$\therefore B = \dfrac{\frac{1}{2} I\, 2\omega \,.\, A\omega}{W}$

or $\quad I = \dfrac{BW}{A\omega^2}$ (9.4)

i.e. I will be the required moment of inertia to maintain the speed of the flywheel to produce a fluctuation of energy of B.

Example 9.4

A single-cylinder four-stroke engine has a mean speed of 250 rev/min. During the power stroke 10 kJ of work is done by the expanding gas on the piston, and the work done by the piston on the gas during compression is 100 J. If the fluctuation of speed is not to exceed 2 per cent above or below the mean, find the moment of inertia of the flywheel required, and the power of the engine. Neglect work done during induction and exhaust.

$$\text{Mean torque of engine} = \frac{\text{Net work done per cycle}}{\text{Total angle turned/cycle}}$$

The total angle turned/cycle, is 4π for a 4-stroke engine

$\therefore$ mean torque $= \dfrac{10\ 000 - 100}{4\pi}$

$\qquad\qquad\qquad = \underline{787 \cdot 8 \text{ Nm}}$

Engine power output $=$ mean torque $\times$ speed

$\qquad\qquad\qquad\qquad = 787 \cdot 8 \times 250 \times \dfrac{2\pi}{60}$

$\qquad\qquad\qquad\qquad = \underline{20 \cdot 6 \text{ kW}}$

Average work done per stroke

$\qquad\qquad\qquad = $ mean torque $\times$ angle turned per stroke
$\qquad\qquad\qquad = 787 \cdot 8 \times \pi$
$\qquad\qquad\qquad = \underline{2475 \text{ watts}}$

Excess work done during power stroke $= 10\ 000 - 2475$
$\qquad\qquad\qquad\qquad\qquad\qquad\quad = \underline{7525 \text{ watts}}$

Deficiency of work done during compression stroke

$$= 2475 - 100$$
$$= \underline{2375 \text{ watts}}$$

Fluctuation of energy $= 7525 - 2375 = \underline{5150 \text{ watts}}$

$$= \tfrac{1}{2} I \left(\omega_1^2 - \omega_2^2 \right)$$

A 2 per cent fluctuation of speed gives $\omega_1 = 255$ rev/min, and $\omega_2 = 245$ rev/min

$$\therefore \text{ re-arranging } I = \frac{2 \times 5150}{(255^2 - 245^2) \left(\dfrac{2\pi}{60} \right)^2} = \underline{187 \cdot 9 \text{ kg m}^2}$$

Example 9.5

The turning-moment diagram for a two-stroke reciprocating gas engine shows that the excess energy produced per cycle is 15 kJ, and the speed of the engine is required to fluctuate by no more than 1 per cent of the mean speed of 500 rev/min. Calculate the moment of inertia of the flywheel required, and the coefficient of fluctuation of speed.

$$500 \text{ rev/min} = 500 \times \frac{2\pi}{60} \text{ rad/s}$$
$$= 52 \cdot 36 \text{ rad/s}$$

Maximum speed $= 52 \cdot 36 + 0 \cdot 5236 = 52 \cdot 8836$ rad/s
Minimum speed $= 52 \cdot 36 - 0 \cdot 5236 = 51 \cdot 8364$ rad/s

Change in kinetic energy of flywheel $= \tfrac{1}{2} I (\omega_1^2 - \omega_2^2)$
$$= \tfrac{1}{2} I (52 \cdot 8836^2 - 51 \cdot 8364^2)$$
$$= \underline{54 \cdot 84 \ I \text{ joules}}$$

This energy is obtained from the excess energy per cycle, thus:

$$54 \cdot 84 \ I = 15 \times 10^3$$
$$\underline{I = 273 \cdot 5 \text{ kgm}^2}$$

The coefficient of fluctuation of speed $= \dfrac{\omega_1 - \omega_2}{\omega}$

$$= \frac{52 \cdot 8836 - 51 \cdot 8364}{52 \cdot 36} = \underline{0 \cdot 02}$$

Example 9.6

A punching machine is used to stamp out circular discs from sheet metal.

The machine is fitted with a flywheel of radius of gyration $0 \cdot 95$ m which has a speed of 200 rev/min just before the beginning of each stamping. It is calculated that 4 kJ of work are required to stamp out each disc, and this energy is assumed to come solely from the flywheel. If the speed reduction during stamping is not to exceed 20 per cent of the maximum calculate:

a the mass of the flywheel required
b the torque that must be applied to the flywheel in order to regain full speed within 5 seconds.

Let M kg be the mass of the flywheel, then

$$\begin{aligned}
\text{moment of inertia of flywheel} &= M\,k^2 \\
&= M\,0 \cdot 95^2 \\
&= 0 \cdot 903\,M \text{ kg m}^2
\end{aligned}$$

$$\begin{aligned}
\text{Speed just prior to stamping} &= 200 \text{ rev/min} \\
&= 200 \times \frac{2\pi}{60} \text{ rad/s} \\
&= 20 \cdot 9 \text{ rad/s}
\end{aligned}$$

$$\begin{aligned}
\text{Speed just after stamping} &= 0 \cdot 8 \times 20 \cdot 9 \\
&= 16 \cdot 76 \text{ rad/s}
\end{aligned}$$

$$\begin{aligned}
\text{Change in kinetic energy of flywheel} &= \tfrac{1}{2} I \left(\omega_1^2 - \omega_2^2 \right) \\
&= \tfrac{1}{2} I \left(20 \cdot 9^2 - 16 \cdot 76^2 \right) \\
&= 77 \cdot 95 \times I \\
&= 77 \cdot 95 \times 0 \cdot 903\,M \\
&= 70 \cdot 39\,M \text{ joules.}
\end{aligned}$$

$$\begin{aligned}
\text{The change in kinetic energy} &= \text{work done in stamping} \\
70 \cdot 39\,M &= 4000 \\
M &= \underline{56 \cdot 8 \text{ kg}}
\end{aligned}$$

The acceleration α, required to accelerate the flywheel from $16 \cdot 76$ rad/s to $20 \cdot 9$ rad/s in 5 s is given by:

$$\alpha = \frac{\omega_1 - \omega_2}{t}$$

$$= \frac{20 \cdot 9 - 16 \cdot 76}{5} = \underline{0 \cdot 828 \text{ rad/s}^2}$$

$$\begin{aligned}
\text{The moment of inertia of flywheel} &= mk^2 \\
&= 56 \cdot 8 \times 0 \cdot 95^2 \\
&= 51 \cdot 26 \text{ kgm}^2
\end{aligned}$$

$$\begin{aligned}
\text{Thus torque required} &= I\alpha \\
&= 51 \cdot 26 \times 0 \cdot 828 \\
&= \underline{42 \cdot 44 \text{ Nm}}
\end{aligned}$$

Example 9.7

The mean speed of the crankshaft of an engine is 120 rev/min, but during any one revolution the speed varies from a minimum of 118 rev/min to a maximum of 122 rev/min. A flywheel of mass 500 kg and radius of gyration $1 \cdot 2$ m is keyed to the crankshaft. If the work done per cycle is 16 kJ calculate:

a the maximum fluctuations of energy in the flywheel
b the coefficient of fluctuation of speed, and
c the coefficient of fluctuation of energy.

$$\omega_1 = 122 \times \frac{2\pi}{60} = 12 \cdot 78 \text{ rad/s}, \quad \omega_2 = 118 \times \frac{2\pi}{60} = 12 \cdot 36 \text{ rad/s}$$

$$\omega = 120 \times \frac{2\pi}{60} = 12 \cdot 57 \text{ rad/s}$$

a From equation (9.1), the fluctuations of energy, E is given by

$$E = \tfrac{1}{2} I(\omega_1^2 - \omega_2^2)$$

$$= \tfrac{1}{2} \times 500 \times 1 \cdot 2^2 \left[\left(122 \times \frac{2\pi}{60} \right)^2 - \left(118 \times \frac{2\pi}{60} \right)^2 \right]$$

$$= \tfrac{1}{2} \times 500 \times 1 \cdot 2^2 \, (12 \cdot 78^2 - 12 \cdot 36^2)$$

$$= \underline{3985 \cdot 3 \text{ joules}}$$

b the coefficient of fluctuation of speed, A, is given by

$$A = \frac{\omega_1 - \omega_2}{\omega}$$

$$= \frac{12 \cdot 78 - 12 \cdot 36}{12 \cdot 57}$$

$$= \underline{0 \cdot 033}$$

c From equation (9.4),

$$I = \frac{BW}{A\omega^2}$$

or, re-arranging $B = \dfrac{IA\omega^2}{W}$

$$= \frac{500 \times 1 \cdot 2^2 \times 0 \cdot 033 \times 12 \cdot 57^2}{16 \times 10^3}$$

$$= \underline{0 \cdot 235}$$

Guided solution

A machine tool performs a certain operation four times each minute, and each operation takes 10 seconds. The machine is driven by a 3 kW electric motor and during each operation the speed of the machine falls from 400 rev/min to 300 rev/min. A flywheel of radius of gyration $0 \cdot 7$ m is fitted to the shaft of the machine in order to ensure that the speed of the machine is restored to its initial operating speed of 400 rev/min at the beginning of each operation. Calculate the mass of the flywheel required, and the amount of energy expended during each operation.

Step 1.   Let m kg be the mass of the flywheel required, and thus calculate the moment of inertia of the flywheel in terms of m. $(0 \cdot 49 \ m \ \mathrm{kgm}^2)$

Step 2.   Convert the maximum and minimum speeds into radians per second, ($42 \cdot 9$ and $31 \cdot 4$ rad/s respectively) and calculate the energy expended during an operation again in terms of m. $(188 \cdot 56 \ m \ \mathrm{joules})$

Step 3.   Calculate the time that exists between each operation for the flywheel to regain the amount of energy expended in step 2, and thus the amount of energy per second that the motor must supply to the flywheel. $(37 \cdot 7 \ m \ \mathrm{joules/s})$

Step 4.   Equate the answer to step 3 to the power output of the motor, and hence evaluate m, the mass the flywheel required. $(79 \cdot 6 \ \mathrm{kg})$

Step 5.   Substitute the mass into the answer to step 2 to find the energy expended during each operation. $(15 \ \mathrm{kJ})$

Problems 9.2

1   A flywheel of radius of gyration $1 \cdot 2$ m gives up to 10 kJ of energy as its speed falls from 200 rev/min to 180 rev/min. Determine its mass.

2   The crankshaft of an engine has a mean speed of 120 rev/min, but its speed is found to fluctuate between 123 and 117 rev/min in each cycle. A flywheel of mass 500 kg and radius of gyration $1 \cdot 4$ m is keyed to the crankshaft and this helps to even out the fluctuations of energy that occur during a revolution. Calculate the maximum fluctuation of energy and the coefficient of fluctuation of speed.

3   In a four-stroke single-cylinder gas engine the work done during the expansion stroke is 800 joules and the work done during the compression stroke is 55 joules. It is assumed that the work done during induction and exhaust strokes is negligible. Calculate the mean turning moment acting on the crankshaft, and the power of the engine if its average speed is 250 rev/min.

   If the fluctuation of speed is not to exceed ± 2 per cent, find the moment of inertia of the flywheel required.

4   A machine press is driven by a motor which delivers 3 kW continuously through a shaft which is fitted with a flywheel of moment of inertia 750 kgm^2. At the beginning of a pressing operation the machine is rotating at 240 rev/min, and each operation takes two seconds and consumes 10 kJ of energy. Calculate the speed of the flywheel after each operation and the number of operations that can be done in one hour.

9.6   Balancing of rotating masses

An out-of-balance mass rotating about an axis will experience a centripetal force which will be reacted by a centrifugal force acting on the axis. If the axis is a shaft running on bearings then it will be necessary to compensate for the imbalance to avoid vibration and damage to the bearings (Fig. 9.9).

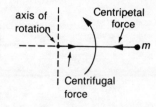

Fig. 9.9

9.7   Static balance in a single plane

Consider the static balance in the same plane of masses m_1, m_2, m_3, m_4 at distances r_1, r_2, r_3, r_4 from a common pivot point O (Fig. 9.10). For static equilibrium of the system the resultant moment of the masses must be zero.

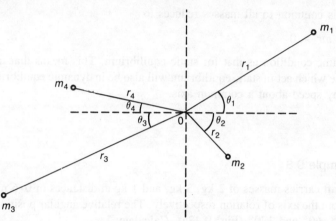

Fig. 9.10

Thus taking moments about O

$$m_1 g r_1 \cos \theta_1 + m_2 g r_2 \cos \theta_2 - m_3 g r_3 \cos \theta_3 - m_4 g r_{4\cos} \theta_4 = 0$$

or simply

$$\Sigma mr \cos \theta = 0$$

since g is common.

Alternatively a vector diagram can be drawn the sides of which represent the vectors $m_1 r_1$, $m_2 r_2$, $m_3 r_3$, $m_4 r_4$ the directions of each being chosen as θ_1, θ_2, θ_3, θ_4 (Fig. 9.11) and taken in order.

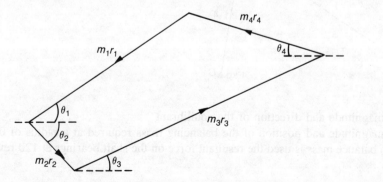

Fig. 9.11

The vector polygon closes if the system is in equilibrium. If the system is not in equilibrium then the vector diagram does not close and the equilibriant (i.e. the vector *mr* that gives equilibrium) is the value of the closing side.

9.8   Dynamic balance in a single plane

For dynamic equilibrium of masses rotating in a single plane the centrifugal forces acting at their centre of rotation are balanced, thus for the masses of Fig. 9.10.

$$\Sigma\ mr\omega^2 = 0$$

which, as ω is common to all masses reduces to

$$\Sigma\ mr = 0$$

This is the same condition as that for static equilibrium. This means that masses in the same plane which act in static equilibrium will also be in dynamic equilibrium when rotating at any speed about a common axis.

Example 9.8

A rotating shaft carries masses of 2 kg, 3 kg, and 1 kg at distances of $0\cdot6$ m, $0\cdot9$ m and $1\cdot4$ m from the axis of rotation respectively. The relative angular positions of the masses are 0°, 60° and 170° (Fig. 9.12a). Calculate:

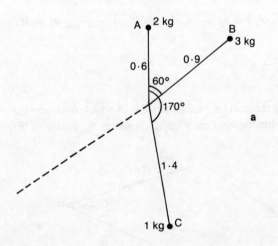

Fig. 9.12

a   the magnitude and direction of the equilibrant
b   the magnitude and position of the balancing mass required at a radius of $0\cdot5$ m
c   if no balance mass is used the resultant force on the shaft bearings at 120 rev/min.

mass m(kg)	radius r(m)	mr
2	0·6	1·2
3	0·9	2·7
1	1·4	1·4

The products of mr are respectively

> 1·2 kg$m$ at 0°
> 2·7 kg$m$ at 60°
> 1·4 kg$m$ at 170°

The vector diagram representing these values is drawn taking each mr value in order and in the direction of the radius from the shaft outwards. From the diagram (Fig. 9.12b):

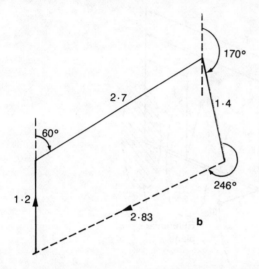

a Vector Mr to give equilibrium is 2·83 kg$m$ at 246°.
b Let the mass required at a radius of 0·5 m be M kg

then $M \times 0·5 = 2·83$

$$M = \frac{2·83}{0·5} = \underline{5·66 \text{ kg}}$$

Thus the balancing mass required at 0·5 m is 5·66 kg at 246°.
c If no balance mass is used the resultant force will be equal and opposite to the equilibriant.

Thus resultant force $= mr\omega^2$

$$= 2·83 \times \left(\frac{120 \times 2\pi}{60} \right)^2 = 446·9 \text{ N}$$

at $246° - 180° = \underline{66°}$ to the direction of 2 kg mass.

9.9   Static balance in more than one plane

The *static* balance of a shaft carrying rotating masses is unaffected by the masses being in different planes, and the analysis is the same as in section 9.7.

9.10   Dynamic balance in more than one plane

Multi-plane balancing is achieved by ensuring that

(a) the sum of the resultant forces is zero, i.e. $\Sigma mr = 0$
(b) the sum of the resultant couples between planes is zero, i.e. $\Sigma mr\ell = 0$ where ℓ is the distance between planes carrying masses, for example, given masses m_1, m_2, m_3 at radii r_1, r_2, r_3 in planes at distances ℓ_1, ℓ_2, ℓ_3 from some reference plane (Fig. 9.13).

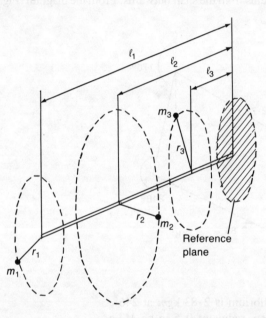

Fig. 9.13

Then: $m_1 r_1 \ell_1 + m_2 r_2 \ell_2 + m_3 r_3 \ell_3 = 0$. The vector polygons which can be drawn that satisfy the above conditions will both be closed. Note that to achieve the second condition, i.e. that the sum of the resultant couples is zero, then an out-of-balance force may require at least *two* added masses in different planes for dynamic balance.

Usually the reference plane is chosen to be the plane of revolution of one of the unknown masses, as this eliminates one unknown couple.

Example 9.9

A rotating shaft carries an out-of-balance mass of 2 kg at a radius, 0·5 m. Calculate the sizes of the balance masses required in planes 0·4 m and 0·6 m on either side of the plane carrying the 2 kg mass, and at radii of 0·3 m and 0·5 m respectively from the shaft axis.

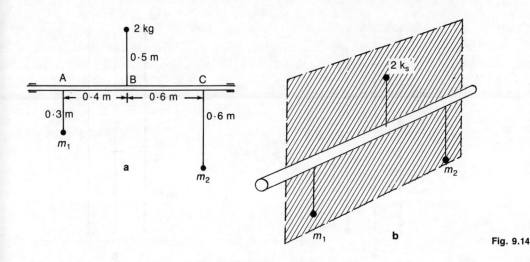

Fig. 9.14

Let the required masses be m_1 and m_2 (Fig. 9.14). For complete dynamic balance there must be no out-of-balance force or couple.

For dynamic balance, taking the reference plane to be through A the plane of revolution of m_1 of the system. The moments of mr values about A

$$\Sigma mr\ell = 0$$
$$m_2 \times 0\cdot6 \times (0\cdot4 + 0\cdot6) - 2 \times 0\cdot5 \times 0\cdot4 = 0$$
$$\therefore\ m_2 = \frac{0\cdot4}{0\cdot6} = \tfrac{2}{3}\,\text{kg}$$

This satisfies the condition *numerically*, but for couples to balance they *must act in the same longitudinal plane*. Therefore m_2 and the 2 kg mass lie in the same plane and are diametrically opposite. Taking moments about C

$$\Sigma mr\ell = 0$$
$$m_1 \times 0\cdot3 \times (0\cdot4 + 0\cdot6) = 2 \times 0\cdot5 \times 0\cdot6$$
$$\therefore\ m_1 = \frac{0\cdot6}{0\cdot3} = 2\,\text{kg}$$

and again the couples will balance providing m_1 lies in the same plane as, and diametrically opposite to, the 2 kg mass. Thus m_1, m_2 and the 2 kg mass all lie in the same longitudinal plane. For static balance there must be no out-of-balance force, i.e. $\Sigma mr = 0$.

Thus

$$\begin{aligned} m_1 r_1 + m_2 r_2 - 2 \times 0\cdot5 \\ = 2 \times 0\cdot3 + \tfrac{2}{3} \times 0\cdot6 - 2 \times 0\cdot5 \\ = 0\cdot6 + 0\cdot4 - 1\cdot0 \\ = 0 \end{aligned}$$

The masses are thus both statically and dynamically balanced.

The vector polygons representing this situation are straight lines. Fig. 9.15a represents the polygon for static balance and Fig. 9.15b that for dynamic balance.

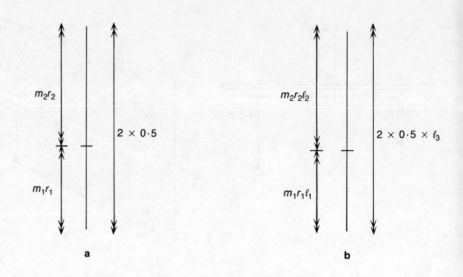

Fig. 9.15

a

b

For the dynamic balance of several masses in different planes of rotation by adding two masses a similar process is used. Each of the given masses is taken in turn and the method of the last example applied. The resultant of the two components is found by completing a parallelogram.

Example 9.10

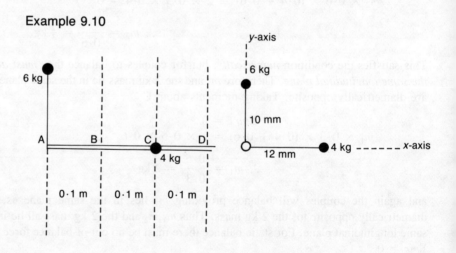

Fig. 9.16

A rotating shaft ABCD carries masses of 6 kg and 4 kg 90° apart in planes A and C and offset from the axis of the shaft by 10 mm and 12 mm respectively. Find the magnitude of two masses, one in plane B and the other in plane D, offset by 15 mm and 20 mm respectively from the axis of the shaft, required to balance the shaft (Fig. 9.16).

Let the two directions of the 6 kg and 4 kg masses be the y and x axes. Let m_1 be the mass required in plane B and m_2 the mass required in plane D.

The 6 kg and 4 kg masses are considered separately. For dynamic balance taking moments about plane D. For the 6 kg mass *only*, we consider an offset force in the same plane as the 6 kg mass and the shaft. This will be the component of m_1 in the y axis, i.e. $(m_1)_y$.

$$(m_1)_y \, r_{1y} \times 0 \cdot 2 - 6 \times 0 \cdot 01 \times 0 \cdot 3 = 0$$

where $(m_1)_y$ is a component of m_1 at radius r_{1y}.

Thus $(m_1)_y \, r_{1y} = 0 \cdot 09$ kg m.

For the 4 kg mass *only*

$$(m_1)_x \times r_{1x} \times 0 \cdot 2 - 4 \times 0 \cdot 012 \times 0 \cdot 1 = 0$$

where $(m_1)_x$ is the other component of m_1 at radius r_{1x}

Thus $(m_1)_x r_{1x} = 0 \cdot 024$ kg m
$$\therefore \ m_1 r_1 = \sqrt{0 \cdot 09^2 + 0 \cdot 024^2} = 0 \cdot 093 \text{ kg m}$$

$$\therefore \ m_1 = \frac{0 \cdot 093}{0 \cdot 015} = 6 \cdot 2 \text{ kg}$$

Figure 9.17 shows this diagramatically.

From the diagram

$$\text{Tan } \theta_1 = \frac{0 \cdot 024}{0 \cdot 09} = 0 \cdot 2667$$

$$\theta_1 = 14° \ 56' \text{ to vertical}$$

For static balance, $\Sigma mr = 0$
for the 6 kg mass only

$$(m_1)_y r_{1y} + (m_2)_y r_{2y} - 6 \times 0 \cdot 01 = 0$$
$$\therefore \ (m_2)_y r_{2y} = 0 \cdot 06 - 0 \cdot 09 = -0 \cdot 03 \text{ kg m}$$
(i.e. in same direction as 6 kg mass)

for 4 kg mass only

$$(m_1)_x r_{1x} + (m_2)_x r_{2x} - 4 \times 0 \cdot 012 = 0$$
$$\therefore \ (m_2)_x r_{2x} = 0 \cdot 048 - 0 \cdot 024 = \underline{0 \cdot 024} \text{ kg m}$$
$$\therefore \ m_2 r_2 = \sqrt{0 \cdot 03^2 + 0 \cdot 024^2}$$
$$= \underline{0 \cdot 0384} \text{ kg m}$$

$$\therefore \ m_2 = \frac{0 \cdot 0384}{0 \cdot 02} = \underline{1 \cdot 92 \text{ kg}}$$

Figure 9.18 shows this diagramatically.

$$\text{Tan } \theta_2 = \frac{0 \cdot 03}{0 \cdot 024}$$

$$= 1 \cdot 25$$
$$\underline{\theta_2 = 51° \ 20'}$$

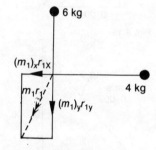

Fig. 9.17   Plane B

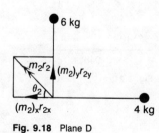

Fig. 9.18   Plane D

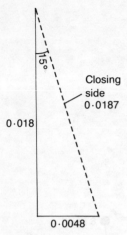

Fig. 9.19

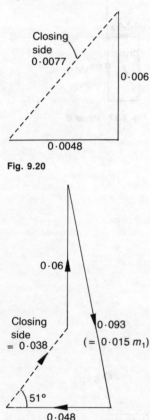

Fig. 9.20

PLANE	m(kg)	r(m)	ℓ(m)	mr	$mr\ell$
A	6	0·01	0·3	0·06	0·018
B	m_1	0·015	0·2	0·15 m_1	0·003 m_1
C	4	0·012	0·1	0·048	0·0048
D	m_2	0·020	0	0·020 m_2	0

Alternatively, taking plane D as the reference plane. The couple polygon can be drawn using the data from the $mr\ell$ column (Fig. 9.19). From the measured closing side

$$0.003\ m_1 = 0.0187$$

$$\therefore\ m_1 = \frac{0.187}{0.003} = \underline{6.2\text{ kg}}$$

at an angle of 15° to vertical.

Taking plane B as the reference plane and noting that vectors on one side of this plane must be regarded as positive and those on the other as being negative and so reversed in direction (Fig. 9.20). From closing side

$$0.004\ m_2 = 0.0077$$

$$m_2 = \underline{1.93\text{ kg}}\text{ at 51° to horizontal}$$

PLANE	m	r	ℓ	mr	$mr\ell$
A	6	0·01	0·1	0·06	0·006
B	m_1	0·015	0	0·015 m_1	0
C	4	0·012	−0·1	0·048	−0·0048
D	m_2	0·02	−0·2	0·02 m_2	−0·004 m_2

Alternatively m_2 can be found by drawing the force polygon using the mr values (noting that the value of m_1 has been found), Fig. 9.21. From the closing side

$$0.02\ m_2 = 0.038$$

$$\therefore\ m_2 = \underline{1.9\text{ kg}}\text{ at 51° to horizontal}$$

9.11   Forces on the bearings due to dynamic imbalance

It has been shown that for a shaft to be in complete dynamic balance there must be no unbalanced force or couple.

In practice if a shaft suffers from dynamic imbalance then the reactions at the bearings of the shaft provide the necessary couple to maintain balance. As the shaft rotates, then the bearing reactions are constantly changing in direction and magnitude. This condition can result in unwanted vibration and bearing wear if it is allowed to persist.

To calculate the forces exerted on the bearings due to dynamic imbalance the product mr required at the bearings to balance the shaft are found. This product multiplied by ω^2 then gives the inertia forces at each bearing. The direction of the inertia forces will be in the opposite direction to the forces necessary to give equilibrium. For complete analysis the normal static reaction will also have to be taken into account.

Fig. 9.21

Example 9.11

A shaft AB rotates in bearings 2 m apart, and carries a rotor of mass 30 kg at a distance of 0·6 m from one end (Fig. 9.22). The centre of gravity of the rotor is offset 2 mm from the axis of the shaft. Calculate the dynamic loading on the bearings at a rotational speed of 1500 rev/min, and the maximum and minimum total load that they bear.

Let m_1r_1 and m_2r_2 be the mr values at A and B respectively.

Plane A reference plane

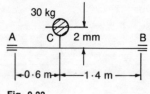

Fig. 9.22

PLANE	m	r	ℓ	mr	$mr\ell$
A	m_1	r_1	0	m_1r_1	0
B	30	0·002	0·6	0·06	0·036
C	m_2	r_2	2·0	m_2r_2	$2\,m_2r_2$

From $mr\ell$ column

$$2m_2r_2 = 0\cdot036$$
$$\therefore\ m_2r_2 = 0\cdot018 \text{ kg m}$$

The dynamic force at B $= m_2r_2\omega^2$

$$= 0\cdot018 \times \left(1500 \times \frac{2\pi}{60}\right)^2$$

$$= 444 \text{ N}$$

For static equilibrium, using the mr column

$$m_1r_1 + m_2r_2 = 0\cdot06$$
$$m_1r_1 = 0\cdot06 - 0\cdot018$$
$$= 0\cdot042 \text{ kg m}$$

The dynamic force at A $= m_1r_1\omega^2$

$$= 0\cdot042 \times \left(1500 \times \frac{2\pi}{60}\right)^2$$

$$= \underline{1036\cdot3 \text{ N}}$$

The force and couple polygons (Fig. 9.23a and b) are both straight lines indicating that the dynamic forces at the bearings are in the same longitudinal plane as the unbalanced load of the rotor.

The normal reactions at the bearings due to the dead weight of the rotor are calculated normally.

Thus, taking moments about B (Fig. 9.24)

$$R_A \times 2 = 30\,g \times 1\cdot4$$
$$\therefore\ R_A = 206\cdot01 \text{ N}$$

and, taking moments about A

$$R_B \times 2 = 30\,g \times 0\cdot6$$
$$\therefore\ R_B = 88\cdot29 \text{ N}$$

Fig. 9.23

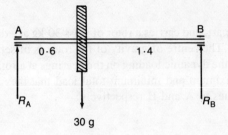

Fig. 9.24

The maximum total loads on the bearings occur when the dynamic and normal reactions occur in the same direction (i.e. both upwards).

$$\text{Maximum total load on A} = 1036 \cdot 3 + 206 \cdot 01$$
$$= \underline{1242 \cdot 31 \text{ N}}$$
$$\text{Maximum total load on B} = 444 + 88 \cdot 29 = \underline{532 \cdot 29 \text{ N}}$$

The minimum total loads on the bearings occur when the dynamic and normal reactions occur in opposite directions.

$$\text{Minimum total load on A} = 1036 \cdot 3 - 206 \cdot 01$$
$$= \underline{830 \cdot 29 \text{ N}}$$
$$\text{Minimum total load on B} = 444 - 88 \cdot 29 = \underline{355 \cdot 71 \text{ N}}$$

Example 9.12

A rotor has a mass of 225 kg and is centrally located on a shaft running in bearings 1·15 m apart. The rotor is balanced statically by adding masses A and B of value 10 kg and 12 kg at radii of 0·375 m and 0·45 m respectively in planes 0·5 m and 0·4 m on either side of the rotor. The masses have an angular displacement of 90°. Calculate the distance of the centre of gravity of the rotor from the axis of rotation and its angular displacement from the mass A. Find also the forces at the bearings at 50 rev/min.

Fig. 9.25 shows the configuration of rotor (C) and masses A and B in relation to the bearings D and E.

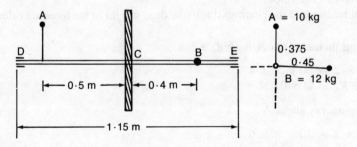

Fig. 9.25

Taking a reference plane through bearing E

Plane	m (kg)	r (m)	l	mr	mrl
A	10	0·375	1·075	3·75	4·031
B	12	0·45	0·175	5·4	0·945
C	225	r_C	0·575	$225\,r_C$	3·674
D	m_D	r_D	1·15	$m_D r_D$	$1·15\,m_D r_D$
E	m_E	r_E	0	$m_E r_E$	0

For static equilibrium the mr values of masses in planes A, B and C form a force polygon which is closed (Fig. 9.26).

The closing side is equivalent to $225\,r_C$

$$\therefore\ 6·4 = 225\,r_C$$
$$\therefore\ r_C = 0·0284 \text{ m}$$
$$= \underline{28·4 \text{ mm}}$$

at $124°\ 47'$ to direction of A.

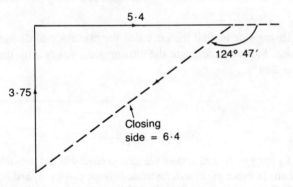

5·4

$124°\ 47'$

3·75

Closing
side = 6·4

Fig. 9.26

For dynamic balance the couple polygon must close. Thus from the mrl column based on a reference plane through E (Fig. 9.27):

The closing side $1·15\,m_D r_D = 2·875$
$$\therefore\ m_D r_D = 2·5$$

The force on the bearing at $D = m_D r_D \omega^2$

$$= 2·5 \times \left(50 \times \frac{2\pi}{60}\right)^2$$

$$= \underline{68·5 \text{ N}}$$

Guided solution

A crankshaft ABCDE has three cranks at B, C and D each of radius 120 mm arranged at 120° to each other and 500 mm apart. The three cranks are equivalent to masses of 20 kg, 30 kg and 40 kg respectively each at the crank radius. Calculate the magnitude of the balance masses required in plane A 200 mm from B at a radius of 200 mm, and in plane E 600 mm from D at a radius of 300 mm.

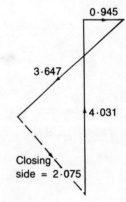

0·945

3·647

4·031

Closing
side = 2·075

Fig. 9.27

Step 1.   Draw the arrangement of the crankshaft ABCDE from a longitudinal and axial point of view indicating the distance between planes and the angles between the given cranks.

Step 2.   Choose a reference plane (say through A) which eliminates one of the unknown couples, and compile a table to give *mr* and *mrl* values. In the table let the unknown masses in the planes A and E be m_A and m_E respectively.

Step 3.   Draw the couple polygon using the values in the *mrl* column of the table, drawing each vector in the direction of each of the masses in planes B, C and D, and ensuring that they follow each other in sequence.

Step 4.   Measure the length of the closing side (the *mr* value) and the angle that it makes with the direction of B. (4·6 kgm at 37·5°).

Step 5.   Equate the *mrl* value obtained to the *mrl* value for plane E to give the required balance mass in plane E (8·52 kg).

Step 6.   Repeat step 2 to step 5 using a reference plane through E. Alternatively draw the force polygon using the column in the original table and using the value of m_E found in step 5.

Step 7.   Measure the closing side to find the *mr* value for plane A and divide by the radius of A to find m_A. Measure the angle the closing side makes with the direction of B (2·75 kg at 239°).

Problems 9.3

1   A rotor of mass 60 kg fitted to a machine shaft has an eccentricity of 3·5 mm relative to the axis of the shaft. In order to achieve complete balance masses m_1 and m_2 are placed at radii of 350 mm in planes of revolution 0·5 mm and 1·2 m from the plane of revolution of the 60 kg mass and on the same side of it. Calculate the magnitude of the masses m_1 and m_2.

2   If, in question 1, the shaft was held in bearings at the planes of the balance masses, calculate the dynamic forces on these bearings without the balance masses and at 150 rev/min.

3   A mass of 75 kg rotates about an axis at a radius of 400 mm. Calculate the magnitude of the balance masses required at radii of 250 mm placed at planes 0·3 m and 0·7 m on either side of the 75 kg mass.

4   A two-cylinder reciprocating engine has a stroke of 0·36 m and the distance between the cylinder centre lines is 0·5 m. At each of the crank pins which are at right-angles to each other, the rotating masses are equivalent to point masses of 200 kg. Calculate the magnitude and angular positions of the balance masses required in planes 1·2 m apart, arranged symmetrically on either side of the cylinders, and at radii of 0·6 m.

5   Masses A, B, C and D are attached to a rotating shaft at radii of 20 mm, 25 mm, 22 mm and 18 mm respectively. The distance between the planes of revolution of A and B is 0·3 m, and of B and C is 0·4 m, and the radius of mass A makes an angle of 90° to that of mass C. If A, C and D are 6 kg, 8 kg and 4 kg respectively calculate:

    a the angles between A, B and D
    b the interplanar distance between C and D
    c the value of mass B
    for complete dynamic balance of the system.

6 Two masses A and B of magnitude 3 kg and 4·5 kg respectively are connected at radii of 0·15 m and 0·2 m respectively to a rotating shaft which is supported in bearings 1 m apart. A and B rotate in planes which are 0·25 m and 0·55 m respectively from one of the bearings. The masses are balanced statically by adding a mass of 5 kg at a radius of 0·15 m at a position on the shaft midway between A and B. Find the angular positions of A and B relative to the balance mass and the dynamic loadings on the bearings at a rotational speed of 200 rev/min.

7 Masses A, B and C of 10 kg, 8 kg and 12 kg respectively are connected to a rotating shaft with mass centres at radii 0·25 m, 0·4 m and 0·15 m respectively from the shaft axis. The interplanar distance between A and B is 0·85 m and between B and C it is 1·25 m. Relative to the radius of A the radii of B and C are angularly displaced by 60° and 140° respectively. Determine the magnitude and angular position with respect to A of two balance masses each at radii of 0·2 m that must be placed in planes midway between A and B, and B and C in order to ensure complete balance.

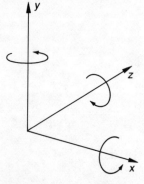

10 Free vibrations

The contents of this chapter will enable you to analyse an undamped freely vibrating system and to solve problems involving such systems with a single degree of freedom.

10.1 Introduction

The subject of vibration was introduced in Chapter 3 with an analysis of Simple Harmonic Motion (SHM). In this chapter the analysis is also confined to vibrations with a single degree of freedom, i.e. a vibrating body is confined to move in one direction only and this may be linear or rotational in nature.

Every object then has a natural frequency of vibration in each of the three co-ordinate axes directions and also in a rotational sense about each of these axes. Thus there are a total of six degrees of freedom or modes of vibration (Fig. 10.1). In order to simplify the mathematical analysis of vibrations most vibrating bodies can be reduced to a simple mass/spring type system. Thus the vibration analysis of a complex system can be reduced to a simple mass with inertia but no elasticity supported by a spring with elasticity but no inertia. The results obtained by using this simplified approach give an acceptable degree of accuracy which can be improved further if allowance is made for the inertia of the spring or elastic structure.

10.2 Simple harmonic motion

An object vibrating freely without damping and with a single degree of freedom has an equation of motion which is reducible to a simple harmonic form. In practice there is always some damping present in virtually all vibration applications and this will cause the vibrations to eventually die out. In previous study we defined SHM as follows:

Any particle that moves so that its acceleration is always directly proportional to its displacement from and directed towards a fixed point in its path is said to move with simple harmonic motion.

The basic equation of SHM is

$$\frac{d^2x}{dt^2} = -\omega_n^2 x \tag{10.1}$$

where x is the displacement from the fixed point, t is the time in seconds, and ω_n is the natural circular frequency. The solution of this second order differential equation is of the form:

Fig. 10.1 The six degrees of freedom

$$x = A \cos \omega_n t + B \sin \omega_n t \tag{10.2}$$

or $x = a \sin (\omega_n t + \alpha)$

where α is the **phase angle** and a is the amplitude of the motion.

If the body begins its motion at the fixed point (i.e. the centre of the motion) then $x = 0$ when $t = 0$ and the solution becomes

$$x = a \sin \omega_n t \tag{10.3}$$

If the body begins its motion at an extreme position then $x = a$ when $t = 0$ and the solution becomes

$$x = a \cos \omega_n t \tag{10.4}$$

where $a =$ amplitude of motion.

Equations (10.3) and (10.4) give identical motion but with a 90° phase shift.

Differentiating equation 10.4 twice with respect to x gives:

$$\frac{dx}{dt} = -a \, \omega_n \sin \omega_n t$$

$$\frac{d^2x}{dt^2} = -a \, \omega_n^2 \cos \omega_n t = -\omega_n^2 x \tag{10.5}$$

The same result is achieved by differentiating equation (10.3) twice.

Two further important relationships are used in SHM. They are the time taken to perform one complete oscillation, termed the periodic time or **period**, T, where

$$T = \frac{2\pi}{\omega_n} \text{ seconds} \tag{10.6}$$

and the number of complete oscillations performed in one second, termed the frequency of the vibration, f where:

$$f = \frac{1}{T} = \frac{\omega_n}{2\pi} \text{ hertz} \tag{10.7}$$

Additional relationships were derived in Chapter 3, namely

$$v = \omega_n \sqrt{a^2 - x^2}$$

where v is the velocity at displacement x and

$$v_{max} = \pm \, \omega_n a$$

where v_{max} is the maximum velocity attained by the body which occurs at the mean position, i.e. when $x = 0$.

Example 10.1

The motion of a body is described by the relationship $2.4 \, d^2x/dt^2 + 300 \, x = 0$, determine the nature of the motion and its period and frequency.

Re-arranging the equation

$$d^2x/dt^2 = -\frac{300}{2\cdot4}x$$

or $$d^2x/dt^2 = -125\,x$$

This equation is of the form $d^2x/dt^2 = -\omega_n^2 x$ and hence the motion is simple harmonic.

In particular $\omega_n^2 = 125$ or $\omega_n = 11\cdot18$ rad/s.

The period $$T = \frac{2\pi}{\omega_n} = \frac{2\pi}{11\cdot18} = 0\cdot56\text{ s}$$

$$f = \frac{1}{T} = \frac{1}{0\cdot56} = \underline{1\cdot78\text{ Hz}}$$

Further examples on SHM can be found in Problems 3.4 and questions 1 and 2 of Problems 10.1.

10.3   The simple mass-spring system

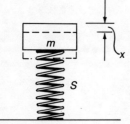

Fig. 10.2

As stated previously a body vibrating naturally with a single degree of freedom can be simplified to a mass-spring system. Consider the mass connected by a spring of stiffness S to a rigid frame or 'ground'. The mass is supposed to be constrained to move only vertically so that there is a single mode of vibration and one natural frequency in this direction. Initially the mass will take up an equilibrium position. Suppose then that the mass is depressed an amount x below the equilibrium position and released. The resulting force on the mass will be upwards and of magnitude Sx. The out-of-balance force will give rise to an acceleration so that applying Newton's Second Law of motion, $P = ma$:

$$Sx = m \times \text{(acceleration)}$$

or $$Sx = -m\frac{d^2x}{dt^2} \qquad (10.8)$$

Note that the acceleration of the mass is given a negative sign since it is in the opposite sense to the displacement.

Re-arrangement of this equation gives

$$\frac{d^2x}{dt^2} + \frac{S}{m}x = 0$$

which is an equation of simple harmonic form with

$$\omega_n = \sqrt{\frac{S}{m}}$$

The solution of the equation is

$$x = a \sin \left(\sqrt{\frac{S}{m}}\, t + \alpha \right)$$

and the natural frequency of vibration,

$$f = \frac{1}{2\pi} \sqrt{\frac{S}{m}}$$

If a more accurate answer is required or if the mass of the spring is significant, the addition of an extra mass to the end equivalent to a third the mass of the spring gives acceptable results.

Thus total end mass = mass + $\frac{1}{3}$ (mass of spring).

Example 10.2

A mass of 2 kg is supported by a spring which hangs vertically from a fixed point. An additional mass of $0 \cdot 5$ kg on the spring produces a further extension of 5 mm. If this additional mass is suddenly removed, find

a the periodic time of the ensuing motion
b the maximum value of velocity and acceleration
c the tension in the spring when the mass is 2 mm from its lowest position.

The stiffness of the spring can be found since the additional mass of $0 \cdot 5$ kg produces an extension of 5 mm, thus

$$S = \frac{0 \cdot 5 \times 9 \cdot 8}{0 \cdot 005} = 981 \text{ N/m}$$

The amplitude of the motion, $a = 5$ mm, the initial displacement.

Applying equation (10.8)

$$981 x = -2 \frac{d^2 x}{dt^2}$$

or, re-arranging $d^2 x/dt^2 = -490 \cdot 5\, x$ (i)
whence $\omega_n = \sqrt{490 \cdot 5} = 22 \cdot 15$ rad/s.
The period,

$$T = \frac{2\pi}{\omega_n} = \frac{2\pi}{22 \cdot 15} = \underline{0 \cdot 284 \text{ s}}$$

From equation (i) the acceleration is a maximum when x is a maximum, i.e. when $x = a$

$$\therefore \left(\frac{d^2 x}{dt^2} \right)_{\max} = -\omega_n^2 a$$

$$= 490 \cdot 5 \times 0 \cdot 005$$
$$= \underline{2 \cdot 45 \text{ m/s}^2}$$

The maximum velocity, $\nu_{max} = \omega_n a$
$$= 22 \cdot 15 \; . \; \times 0 \cdot 005$$
$$= \underline{0 \cdot 11 \text{ m/s}}$$

x is the displacement from the mean position, so that when the mass is 2 mm from its lowest position $x = 3$ mm.

The acceleration when x is 3 mm $= \omega_n^2 \times 0 \cdot 003$
$$= 490 \cdot 5 \times 0 \cdot 003$$
$$= 1 \cdot 47 \text{ m/s}^2 \text{ towards the mean position.}$$

The out-of-balance force causing this acceleration is provided by the difference between the spring tension and the weight, $(T - mg)$ newtons. Thus applying Newton's second law.

$$T - \text{mg} = m \times \text{acceleration}$$
$$T = 2 \, (9 \cdot 81 + 1 \cdot 47) = \underline{22 \cdot 56 \text{ N}}$$

Example 10.3

A valve which moves with simple harmonic motion has a mass of $0 \cdot 4$ kg and a total lift of $5 \cdot 5$ mm. The total time taken to open and close the valve is $0 \cdot 05$ seconds. Determine the accelerating forces which act on the valve stem at the beginning and end of its operation.

Since the total lift is $5 \cdot 5$ mm then the amplitude of the motion

$$= \frac{5 \cdot 5}{2} = \underline{2 \cdot 75 \text{ mm}}$$

The period of the motion $= 0 \cdot 05 \text{ s} = \dfrac{2\pi}{\omega_n}$

$$\therefore \; \omega_n = \frac{2\pi}{0 \cdot 05} = 125 \cdot 66 \text{ rad/s}$$

The acceleration at the beginning and end of the valve operation, i.e. the acceleration at maximum amplitude

$$= \omega_n^2 a$$
$$= (125 \cdot 66)^2 \times 0 \cdot 00275$$
$$= 43 \cdot 42 \text{ m/s}^2$$

The forces that exist at maximum amplitude to cause this acceleration are found by applying Newton's second law, thus

$$p = ma$$
$$= 0 \cdot 4 \times 43 \cdot 42$$
$$= \underline{17 \cdot 37 \text{ newtons}}$$

Now try questions 3 to 6 of Problems 10.1.

10.4   Torsional vibrations

The treatment of a torsional vibrating system is similar to that for the longitudinal mass/spring system. Consider a rotor of moment of inertia I connected via a shaft of torsional stiffness S, to a rigid frame. Suppose that the rotor was given an angular displacement θ from its rest position and then released. The restoring torque on the rotor $= S\theta$, and this out-of-balance torque will give rise to an angular acceleration. Applying Newton's second law for angular motion $T = I\alpha$ to the system gives

$$S\theta = I\left(-\frac{d^2\theta}{dt^2}\right)$$

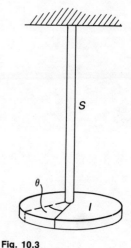

Fig. 10.3

Again the acceleration is negative, being in the opposite sense to the displacement, re-arrangement gives

$$\frac{d^2\theta}{dt^2} + \frac{S}{I}\theta = 0 \tag{10.9}$$

which is an equation of simple harmonic form with $\omega_n = \sqrt{\dfrac{S}{I}}$
The solution of the equation is

$$\theta = \theta_0 \sin\left(\sqrt{\frac{S}{I}}\, t + \alpha\right) \tag{10.10}$$

when θ_0 is the amplitude and the natural frequency,

$$f = \frac{1}{2\pi}\sqrt{\frac{S}{I}} \tag{10.11}$$

The torsional stiffness of a shaft is defined as the amount of torque required to give unit angle of twist. From the torsion of shafts formula

$$S = \frac{T}{\theta} = \frac{GJ}{l} \tag{10.12}$$

where G is the modulus of rigidity.
   $\qquad J$ is the polar second moment of area
and $\quad l$ is the length of the shaft.
Thus the natural frequency of vibration becomes

$$f = \frac{1}{2\pi}\sqrt{\frac{GJ}{Il}} \tag{10.13}$$

As with the mass-spring system if the inertia of the shaft is significant then an allowance can be made by adding one third the inertia of the shaft to the inertia of the rotor. Thus, total inertia of rotor $= I + \frac{1}{3}$ (inertia of shaft).

Example 10.4

A shaft $1\cdot 2$ m long and diameter 3 mm has one end rigidly fixed to a frame support and the other carries a rotor of moment of inertia $0\cdot 034$ kgm^2. If the period of free vibration of the system is $1 \cdot 5$ seconds determine the modulus of rigidity of the material of the shaft.

If the amplitude of the oscillation is $10°$ determine the maximum values of the velocity and acceleration.

$$J \text{ for the shaft} = \frac{\pi D^4}{32} = \frac{\pi (0 \cdot 003)^4}{32} = 7 \cdot 95 \times 10^{12} \ m^4$$

The period $\quad T = \dfrac{1}{f} = 2\pi \sqrt{\dfrac{Il}{GJ}}$

re-arranging $\quad G = \dfrac{4\pi^2 Il}{T^2 J}$

$$= \frac{4\pi^2 \times 1 \cdot 2 \times 0 \cdot 034}{1 \cdot 5^2 \times 7 \cdot 95 \times 10^{-12}} = 90 \cdot 05 \ GN/m^2$$

the amplitude $\theta_0 = 10° = 0 \cdot 1745$ radians.

The period $\quad T = \dfrac{2\pi}{\omega_n}$

$$\therefore \ \omega_n = \frac{2\pi}{T} = \frac{2\pi}{1 \cdot 5} = 4 \cdot 19 \ rad/s.$$

The maximum velocity $= \theta_0 \omega_n$
$$= 0 \cdot 1745 \times 4 \cdot 19$$
$$= \underline{0 \cdot 73 \ rad/s}$$

The maximum acceleration $= \omega_n^2 \theta_0$
$$= 4 \cdot 19^2 \times 0 \cdot 1745$$
$$= 3 \cdot 06 \ rad/s^2.$$

Now try questions 7 to 10 of Problems 10.1.

10.5 Simple and compound pendulums

A simple pendulum is considered to be a concentrated mass at the end of a light string or rod whose mass is insignificant (Fig. 10.4). If the mass is displaced so that the string is inclined at an angle θ to the vertical and released it will perform simple harmonic oscillations about the mean position. The moment of the out-of-balance force about the pivot point is $mgl \sin \theta$, and so applying Newton's second law for angular motion to the system.

$$mgl \sin \theta = -I \frac{d^2\theta}{dt^2}$$

Where I is the moment of inertia of the mass about the pivot point $= ml^2$, also if θ is small then $\sin \theta \simeq \theta$ in radians.

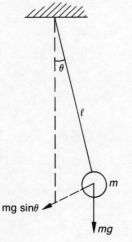

Fig. 10.4

mg sinθ

θ

ℓ

m

mg

Then $mg \, l\theta = -ml^2 \dfrac{d^2\theta}{dt^2}$

or $\dfrac{d^2\theta}{dt^2} = -\dfrac{g}{l}\theta$ (10.14)

which is an equation of SHM form with $\omega_n = \sqrt{\dfrac{g}{l}}$

Thus the period T of the motion is given by

$$T = 2\pi \sqrt{\dfrac{l}{g}}$$ (10.15)

Fig. 10.5

In a compound pendulum the mass is not considered concentrated at a point but at a centre of gravity G a distance x from the pivot point (Fig. 10.5). Again the moment of the out-of-balance force about the pivot assuming θ is small, is

$mg \, x \, \theta$

Applying Newton's second law

$$mg \, x \, \theta = -I \dfrac{d^2\theta}{dt^2}$$

or $\dfrac{d^2\theta}{dt^2} = -\dfrac{mgx}{I}\theta$ (10.16)

Again this is an equation of SHM form and thus the frequency

$$f = \dfrac{1}{2\pi} \sqrt{\dfrac{mgx}{I}}$$ (10.17)

I is the moment of inertia about the pivot, and if I_G is the moment of inertia about the centre of gravity G, then by the parallel axis theorem

$$\begin{aligned} I &= I_G + mx^2 \\ &= mk^2 + mx^2 \\ &= m(k^2 + x^2) \end{aligned}$$

where k is the radius of gyration of the body about its centre of gravity.

Thus the frequency of oscillation may be expressed in the form

$$f = \dfrac{1}{2\pi} \sqrt{\dfrac{gx}{k^2 + x^2}}$$ (10.18)

Equation (10.18) provides a simple way of determining the radius of gyration of an object. All that is required is to suspend the object from a point and time a number of small oscillations to determine the frequency.

Example 10.5

The connecting rod shown is supported on a knife-edge on the inside of the small-end bearing. The mass of the rod is 2 kg, and the periodic time of its oscillation about the knife-edge is found to be 1·2 seconds. Determine the radius of gyration of the rod about the centre of gravity, G, and the period if the rod were hung from its big-end bearing.

The distance of the knife-edge pivot from G = 212·5 mm = 0·2125 m.

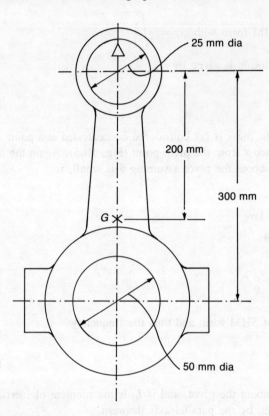

Using equation (10.18).

$$f = \frac{1}{T} = \frac{1}{2\pi} \sqrt{\frac{gx}{k^2 + x^2}}$$

re-arranging $\sqrt{k^2 + x^2} = \dfrac{T}{2\pi} \sqrt{gx}$

$$k^2 + x^2 = \frac{T^2}{4\pi^2} gx$$

$$k = \sqrt{\frac{T^2 gx}{4\pi^2} - x^2}$$

$$= \sqrt{\frac{1·2^2 \times 9·81 \times 0·2125}{4\pi^2} - 0·2125^2}$$

$$= \sqrt{0 \cdot 076 - 0 \cdot 0452}$$
$$= \underline{0 \cdot 176 \text{ m}}$$

If the connecting rod is supported by its big-end bearing, the distance from the knife-edge pivot to

$$G = 300 - 200 + 25 - 125 \text{ mm} = 0 \cdot 125 \text{ m}$$

$$\text{The period } T = \frac{1}{f} = 2\pi \sqrt{\frac{k^2 + x^2}{gx}}$$

$$= 2\pi \sqrt{\frac{0 \cdot 176^2 + 0 \cdot 125^2}{9 \cdot 81 \times 0 \cdot 125}}$$

$$= \underline{1 \cdot 225 \text{ seconds}}$$

Now try questions 11 and 12 of Problems 10.1.

10.6   Transverse vibrations

A bar or beam will also provide an elastic stiffness in a transverse direction i.e. at right-angles to the axis of the beam. For example consider the simply supported beam AB of length l shown (Fig. 10.6) with a point load W at the centre of its span. If the

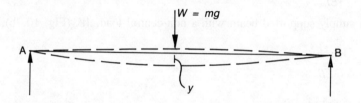

Fig. 10.6

load is depressed from the equilibrium position by a distance y and released the beam will oscillate about its mean position. The load that would be required to produce the deflection y is given by the standard case formula of Fig. 7.3, viz.

$$\frac{48 \, EI \, y}{l^3}$$

and this is the value of the restoring force due to the stiffness of the beam.
   Thus applying Newton's second law, $P = ma$

$$\text{restoring force} = \text{mass} \times \text{acceleration}$$

$$\frac{48 \, EI \, y}{l^3} = - \, m \, \frac{d^2y}{dt^2}$$

$$\text{or} \quad d^2y/dt^2 = - \, \frac{48 \, EI \, y}{ml^3} \tag{10.19}$$

This is an equation of simple harmonic type where

$$\omega_n = \sqrt{\frac{48\,EI}{ml^3}} \,,$$

and the frequency of vibration, f is given by the relationship

$$f = \frac{1}{2\pi} \sqrt{\frac{48\,EI}{ml^3}} \tag{10.20}$$

The static deflection δ under the central load is

$$\frac{Wl^3}{48\,EI} = \frac{mgl^3}{48\,EI}$$

Substituting into (10.20) gives $\quad f = \dfrac{1}{2\pi} \sqrt{\dfrac{g}{\delta}} \tag{10.21}$

Equation (10.21) is universal for a beam carrying a single point load. The expression used for δ must be that appropriate for the load position and the type of support used. Reference to the table of Fig. 7.3 will give the more common expressions. For example for a cantilever of length l with a point load W at the free end (Fig. 10.7a)

$$\delta = \frac{Wl^3}{3EI} \,,$$

whilst for a simply supported beam with a non-central load, W, (Fig. 10.7b),

$$\delta = \frac{Wa^2b^2}{3EIl}$$

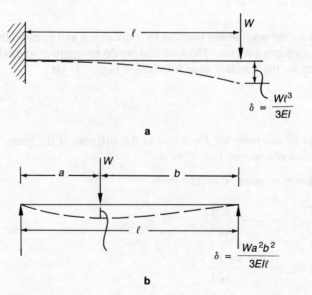

$$\delta = \frac{Wl^3}{3EI}$$

a

$$\delta = \frac{Wa^2b^2}{3EIl}$$

Fig. 10.7   b

Example 10.6

A steel strip 200 mm long acts as a horizontal cantilever and carries a point load of 100 newtons at its free end. The strip is 25 mm wide and 5 mm thick and $E = 200\ \text{GN/m}^2$. Neglecting the mass of the beam, determine the frequency of natural transverse vibrations.

The static deflection δ of the free end of the cantilever

$$= \frac{Wl^3}{3EI}$$

$$= \frac{100 \times 0\cdot2^3 \times 12}{3 \times 200 \times 10^9 \times 0\cdot025 \times 0\cdot005^3} = \underline{5\cdot12 \times 10^{-3}\ \text{m}}$$

The frequency of vibration, $f = \dfrac{1}{2\pi}\sqrt{\dfrac{g}{\delta}}$

$$= \frac{1}{2\pi}\sqrt{\frac{9\cdot81}{5\cdot12 \times 10^{-3}}}$$

$$= \underline{6\cdot97\ \text{Hz}}$$

Now try questions 13 and 14 of Problems 10.1.

10.7   Dunkerley's method for more than one point load

The natural frequency of a beam subjected to several point loads can be found approximately using Dunkerley's Method. If f is the natural frequency of the fully loaded beam; f_0 the natural frequency of the beam due to its own weight; f_1, the natural frequency of the beam due to point load number one, f_2 the natural frequency of the beam due to point load number two and so on, then

$$\frac{1}{f^2} = \frac{1}{f_0^2} + \frac{1}{f_1^2} + \frac{1}{f_2^2} + \ldots \text{ etc.} \tag{10.22}$$

Example 10.7

The simply supported beam shown is 8 m long and subjected to point loads of 30 kN and 20 kN at distances of 2 m and 4 m respectively from one end. Neglecting the mass of the beam, determine the natural frequency of vibration of the system given that $EI = 20\ \text{MNm}^2$.

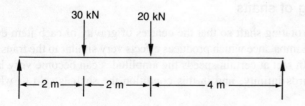

The deflection under the 20 kN load, $\delta_{20} = \dfrac{Wl^3}{48\,EI}$

$$= \frac{20 \times 10^3 \times 8^3}{48 \times 20 \times 10^6}$$

$$= \underline{0\cdot0107 \text{ m}}$$

The natural frequency due to the 20 kN only

$$= f_{20} = \frac{1}{2\pi}\sqrt{\frac{9\cdot81}{0\cdot0107}}$$

$$= \underline{4\cdot82 \text{ Hz}}$$

The deflection under the 30 kN load

due to this load only, $\delta_{30} = \dfrac{Wa^2b^2}{3EIl}$

$$= \frac{30 \times 10^3 \times 2^2 \times 6^2}{3 \times 20 \times 10^6 \times 8}$$

$$= \underline{0\cdot009 \text{ m}}$$

The natural frequency due to 30 kN only

$$= f_{30} = \frac{1}{2\pi}\sqrt{\frac{9\cdot81}{0\cdot009}}$$

$$= \underline{5\cdot255 \text{ Hz}}$$

Neglecting the mass of the beam, $f_0 = 0$

Applying Dunkerley's formula

$$\frac{1}{f^2} = \frac{1}{f_{30}^2} + \frac{1}{f_{20}^2}$$

$$= 0\cdot0792$$

Whence $\underline{f = 3\cdot55 \text{ Hz}}$

Now try questions 15 and 16 of Problems 10.1.

10.8   The whirling of shafts

A rotor mounted on a rotating shaft so that the centres of gravity of each item do not coincide experiences an imbalance which produces effects very similar to the transverse vibration of the shaft. In fact at certain speeds the amplitudes can become very large, in theory tending towards infinity, and in this condition the shaft is said to **whirl**.

Consider the simply supported shaft AB (Fig. 10.8) on rigid bearings and carrying a rotor of mass m kg which has its centre of gravity displaced by an amount e from the axis of the shaft, i.e. the shaft and rotor have axes e apart.

In the stationary mode the mass of the rotor will cause a static deflection δ of the shaft whose mass we will assume is negligible.

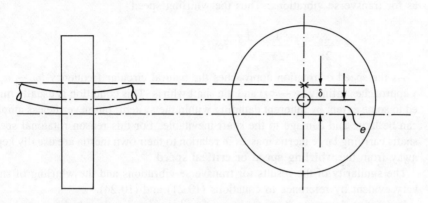

Fig. 10.8

If the rotor is then given an angular velocity ω about the shaft axis then it will experience a centripetal force of value $m(y + e)\omega^2$ where y is the displacement due to the imbalance. This centripetal force is provided by the restoring force due to the stiffness of the beam and is equal to Sy where S is the stiffness. The value of the stiffness will depend on the position of the rotor and the nature of the end fixing in a manner identical to that used for the transverse vibrations of a beam. For example if the rotor were centrally placed on a simply supported shaft then

$$S = \frac{48 \, EI}{l^3}$$

Equating the two expressions for the centripetal force gives:

$$m \, (y + e) \, \omega^2 = Sy$$

or $$(y + e) \, \omega^2 = \frac{S}{m} \, y$$

As before $\dfrac{S}{m} = \omega_n^2$, where ω_n = natural circular frequency

$$\therefore (y + e) \, \omega^2 = \omega_n^2 \, y$$

re-arranging $$y = \pm \, \frac{e}{\dfrac{\omega_n^2}{\omega^2} - 1} \tag{10.23}$$

The $\pm$ sign signifies that y may be of opposite sign to e this depending on whether ω is greater or less than ω_n.

The whirling speed is thus defined as being equal to the natural circular frequency of the shaft, where ω_n has already been defined as

$$\omega_n = \sqrt{\frac{S}{m}} = \sqrt{\frac{g}{\delta}}$$

as for transverse vibrations. Thus the whirling speed

$$= \frac{\omega_n}{2\pi} = \frac{1}{2\pi} \sqrt{\frac{g}{\delta}} \text{ rev/s} \qquad (10.24)$$

As the speed of rotation approaches the natural circular frequency ($\omega \rightarrow \omega_n$) then y approaches infinity ($y \rightarrow \infty$) and the shaft whirls. This condition is usually minimised to some extent by inherent damping within the system, but nevertheless amplitudes can be large and damage to the shaft inevitable. For this reason rotatonal speeds of shafts carrying large inertia masses in relation to their own inertia are usually kept well away from the **whirling speed** or **critical speed**.

The similarity of the results for transverse vibrations and the whirling of shafts is very evident by reference to equations (10.21) and (10.24).

In practice, however, it is not always possible to predict the whirling speed of shafts by reference to the natural frequency of transverse vibration.

Several factors make such comparisons unreliable including inertia torques due to the inclination of the rotor to the axis of rotation, gyroscopic couples due to the motion of the shaft in a direction perpendicular to the axis of rotation, and the inevitable flexibility of the support bearings.

Example 10.8

A light shaft, 10 mm diameter is simply supported on bearings $1 \cdot 0$ m apart and carries a rotor of mass 5 kg at the centre of its span. The rotor axis is eccentric to the shaft axis by $0 \cdot 02$ mm.

Calculate **a** the critical speed at which whirling takes place

        b the maximum deflection of the shaft from the mean position at a speed of 600 rev/min.

The static deflection $\delta = \dfrac{Wl^3}{48\,EI} - \dfrac{5 \times 1^3}{48 \times 200 \times 10^9 \times \dfrac{\pi \times 0 \cdot 01^4}{64}}$

$$= \underline{1 \cdot 06 \times 10^{-3} \text{ m}}$$

The whirling speed is equal to the natural circular

frequency $\omega_n = \sqrt{\dfrac{g}{\delta}} = \sqrt{\dfrac{9 \cdot 81}{1 \cdot 06 \times 10^{-3}}} = 96 \cdot 2 \text{ rad/s}$

$$\therefore \text{ whirling speed } = \frac{96 \cdot 2}{2\pi} \times 60 \qquad = \underline{918 \text{ rev/min}}$$

The deflection $y = \dfrac{e}{\dfrac{\omega_n^2}{\omega^2} - 1}$

$$= \dfrac{0 \cdot 02}{\dfrac{918^2}{600^2} - 1} = \underline{0 \cdot 015 \text{ mm}}$$

Now try questions 17 and 18 of Problems 10.1.

Guided solution

a A rotor of mass 100 kg and diameter $0 \cdot 8$ m is connected to the mid-point of a steel shaft 25 mm diameter and $1 \cdot 2$ m long. Neglecting the mass of the shaft, calculate the natural frequencies of free vibrations in the following modes.

  i longitudinal
 ii transverse
iii torsional

b If the rotor and shaft axes have an eccentricity of $0 \cdot 01$ mm calculate the whirling speed of the shaft and the maximum deflection from the mean position at a speed of 500 rev/min.

i   *longitudinal mode*
Step 1. Calculate the stiffness in the longitudinal direction given that stiffness is defined as the load required to produce unit extension. $(8 \cdot 18 \times 10^7 \text{ N/m})$

Step 2. Calculate using the standard equation the natural frequency of longitudinal vibrations.   (144 Hz)

ii   *transverse mode*
Step 3. Calculate the stiffness in the transverse mode using the standard results for beam deflections. Remember that I in the formula is the second moment of area about a diameter of the shaft.   (106 526 N/m)

Step 4. Using the standard formula calculate the frequency of transverse vibrations. $(5 \cdot 19 \text{ Hz})$

iii   *torsional mode*
Step 5. Calculate the torsional stiffness of the shaft.   (2461 Nm/rad)

Step 6. Calculate the moment of inertia of the rotor given that

$$I = \frac{mr^2}{2} \qquad\qquad (8 \text{ kgm}^2)$$

Step 7. Substituting these values into the standard formula calculate the natural frequency of torsional vibrations. $(2 \cdot 8 \text{ Hz})$

Step 8. The whirling speed is equal to the natural frequency of transverse vibrations. Give the answer in rev/min.   (311 rev/min)

Step 9. The maximum deflection from the equilibrium position can be found using equation (10.23). (0·016 mm)

Problems 10.1

1 A body of mass 50 kg oscillates in a horizontal straight line with simple harmonic motion. The amplitude of the oscillation is 1·3 m and the period is 10 s. Determine:
   a the maximum force acting on the body
   b the velocity of the body when it is 0·3 m from the centre of oscillation
   c the time taken to travel 0·3 m from an extreme point.

2 A valve of mass 8 kg moves horizontally with SHM over a stroke of 0·15 m at the rate of 4 double strokes (i.e. backwards and forwards) per second — events controlled by the valve occur when it is 40 mm from its mid-position and moving towards it, and again when it is 25 mm beyond the mid-position on the same stroke. Determine:
   a the time between these events
   b the maximum force to drive the valve (ignore frictional resistance).

3 A mass of 20 kg is suspended from a spring of stiffness 10 kN/m and is made to vibrate freely with an amplitude of 10 mm. Find
   a the periodic time, the velocity and acceleration when the displacement from the equilibrium position is 8 mm and
   b the time taken to move from this position to the maximum displacement.

4 A mass of 5 kg is hung from the end of a thin wire which is observed to stretch 2·6 mm. Calculate the period of free longitudinal vibration when a mass of 10 kg is hung on the wire.

5 A machine of mass 100 kg is supported by a structure having a stiffness of 10^4 N/m. If the machine is displaced 40 mm from the rest position and then released calculate the natural frequency of free vibrations and the displacement, velocity and acceleration after 1 second.

6 A spring of stiffness 250 N/m has a mass of 1 kg. A mass of 8 kg is attached to the free end and the system set vibrating. Determine the natural frequency of free vibration
   a neglecting the mass of the spring
   b including the mass of the spring.

7 A disc is attached to a mild steel shaft of 5 mm diameter and 1 m long. The shaft is mounted vertically with its upper end fixed and the disc is attached to the lower end. If the moment of inertia of the disc is 0·02 kgm^2, calculate the period of free torsional oscillations. $G = 80$ GN/m^2.

8 If a second disc is now attached to the disc in Q.7, the time for 20 complete torsional oscillations is found to be 2·1 minutes. If the second disc has a mass of 50 kg determine its radius of gyration.

9 The torsional stiffness of a thin shaft is 50 Nm per radian. Such a shaft is used to suspend a solid cylindrical rotor of diameter 0·5 m and when given small torsional vibrations the period is found to be 1·5 seconds. Calculate the mass of the rotor.

10 A shaft 1 m long and 20 mm diameter has one end fixed co-axially with a solid cylinder of diameter 100 mm and thickness 100 mm. The other end of the shaft is rigidly fixed to a support and the whole system is made to perform small torsional vibrations. If the shaft and the cylinder are made of steel of density 8000 kg/m^2 calculate the natural frequency of torsional vibrations accounting for the inertia of the shaft.

11 A connecting rod of mass 20 kg is suspended from a knife-edge from the inner surface of the small end bearing, and is found to perform 50 complete oscillations in 1 minute. The rod is then inverted and suspended from the inner surface of the big end bearing where it made 58 complete oscillations in one minute. The length of the rod between support points is 0·4 m. Calculate the moment of inertia of the rod about an axis through its centre of gravity.

12 A gear wheel is mounted on a knife-edge so that it is able to swing in a vertical plane with its axis horizontal. The knife-edge is 300 mm from the centre of gravity of the wheel, and it is found that the wheel performs 50 complete oscillations in 110 seconds. Calculate the radius of gyration of the wheel about its centre of gravity.

13 A round bar 15 mm diameter and 300 mm long acts as a horizontal cantilever and carries a point load of 150 N at its free end. Neglecting the mass of the beam determine the natural frequency of transverse vibrations.

$E = 200$ GN/m^2

14 A horizontal shaft 25 mm diameter is simply supported between bearings 2 m apart. The shaft carries a point load of 1 kN at a distance of 0·4 m from one bearing. Determine the frequency of free transverse vibrations. $E = 200$ GN/m^2

15 A cantilever beam 4 m long carries point loads of 15 kN and 25 kN at the free end and at mid-span respectively. Neglecting the mass of the beam determine the natural frequency of vibration of the system given that $EI = 20$ MN m^2.

16 A simply supported beam 3 m long carries equal point loads at 1 m and 2 m. The natural frequency of transverse vibrations of the system is found to be 22·4 hertz. If $EI = 20$ MN/m^2 calculate the magnitude of the point loads.

17 A light shaft carries a rotor whose axis is 0·1 mm from the axis of the shaft. The weight of the rotor causes a static deflection of the shaft of 1 mm. Calculate the whirling speed of the shaft. At what speed would the maximum deflection from the mean reach 4 mm?

18 A shaft of diameter 25 mm is supported in bearings 0·5 m apart. At the centre of span is a rotor of mass 60 kg whose centre of gravity is 0·02 mm eccentric to the axis of the shaft. Determine:
a the critical speed of rotation
b the maximum deflection of the shaft from its equilibrium position at a speed of 2000 rev/min. $E = 200$ GN/m^2

Appendix 1

The position of the centroid of a lamina and the first moment of area

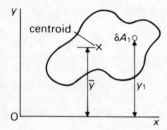

Given a plane lamina of area A in rectangular co-ordinates xy. Let $\bar{y}$ be the distance of the centroid of the lamina from Ox. Consider a small element of the area, δA_1 at a distance y_1 from Ox.

The 'moment' of this area about Ox is $\delta A_1 \times y_1$. To be precise the product $\delta A_1 \times y_1$ is termed the **first moment of area** of the element δA_1 about Ox.

The total area A is comprised of many small elements, δA_1, δA_2, δA_3, etc., at distance y_1, y_2, y_3 etc. from Ox.

The *total* first moment of area of the whole lamina is $A\bar{y}$, and this is equal to the *sum* of the first moments of area of all the elemental areas. Thus

$$A\bar{y} = \delta A_1 y_1 + \delta A_2 y_2 + \delta A_3 y_3 + \ldots \text{ etc.}$$

In the limit, as δA is made infinitesimally small, the sum of all first moments of area of all the elemental areas may be written as an integral, $\int y\, dA$, where the integral sign means 'sum of'.

Thus, *the total first moment of area of the lamina is $\int y\, dA$*, i.e.

$$A\bar{y} = \int y\, dA \quad \text{or} \quad \bar{y} = \frac{\int y\, dA}{A}$$

This formula will thus give the distance of the centroid for the neutral axis of a lamina (or section) from any designated datum. For example let us consider a plane rectangular lamina (or the rectangular section of a beam).

The position of the centroid of a rectangular lamina

Given a rectangular lamina breadth b and depth d, so that $A = bd$. It is required to find the position of the centroid relative to a datum XX along one edge of the beam. We require to evaluate the first moment of area of the beam $\int y\, dA$, and for this we need y as a function of the area A. We do this by dividing the rectangle into thin elemental strips. Consider one of these strips, breadth b, thickness δy and distance y from XX.

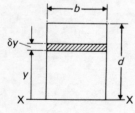

Area of strip $= b \times \delta y$.

First moment of area of strip about XX $= b\delta y \times y$.

Sum of first moments of area of all such elemental strips from $y = 0$ to $y = d$ is $\int_0^d$ by dy. Thus

$$A\bar{y} = \int_0^d by \, dy = \left[\frac{by^2}{2}\right]_0^d = \frac{bd^2}{2}$$

but $A = bd$, therefore $\bar{y} = \frac{bd^2/2}{bd} = \frac{d}{2}$

which, of course, is the result that we should have expected.

Appendix 2

The second moment of area of a lamina

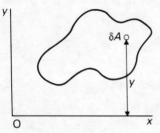

Consider again the lamina first introduced in Appendix 1 and an elemental area δA.
The *first* moment of area of δA about Ox is $\delta A \times y$.
The *second* moment of area of δA about Ox is then $\delta A \times y \times y$ or $y^2 \delta A$.

The *sum* of the second moments of areas of all such elemental area is thus given by $\int y^2 \, dA$, and this is equal to the total second moment of area of the lamina about Ox, I_{Ox}. Thus

$$I_{Ox} = \int y^2 \, dA$$

Again it is useful to apply the theory to actual shapes.

The second moment of area of a rectangle about one edge

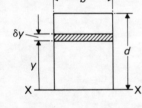

Given the rectangle breadth b, depth d. We require to find I_{XX}, where XX is a datum along the lower edge of the rectangle. Consider an elemental strip breadth b, thickness δy, distance y from XX.

Area of strip $= b \times \delta y$.
First moment of area of strip about XX $= b \times \delta y \times y$.
Second moment of area of strip about XX $= b \times \delta y \times y^2$.
Total second moment of area of all such elemental strips about XX from $y = 0$ to $y = d$ is:

$$\int_0^d by^2 \, dy$$

$$= b \left[\frac{y^2}{3} \right]_0^d = \frac{bd^3}{3}$$

Thus $I_{XX} = \dfrac{bd^3}{3}$

The second moment of area of a rectangle about an axis through its centroid (neutral axis)

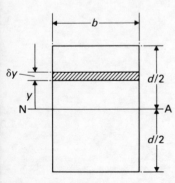

Given a rectangle breadth b, depth d, with the datum axis NA as shown. Consider an elemental strip, breadth b, thickness δy, distance y from NA.

Area of strip $= b \times \delta y$.

First moment of area of strip about NA $= b \times \delta y \times y$.

Second moment of area of strip about NA $= b \times \delta y \times y^2$.

Total second moment of area of all such elemental strips about NA from $y = -d/2$ to $y = +d/2$ is:

$$\int_{-d/2}^{+d/2} by^2 \, dy$$

$$= b \left[\frac{y^3}{3} \right]_{-d/2}^{+d/2}$$

$$= b \left\{ \left[\frac{(d/2)^3}{3} \right] - \left[\frac{(-d/2)^3}{3} \right] \right\}$$

$$= \frac{bd^3}{24} + \frac{bd^3}{24}$$

$$= \frac{bd^3}{12}$$

Thus $I_{NA} = \dfrac{bd^3}{12}$

The parallel axis theorem

If I_{NA} is the second moment of area of a lamina about the neutral axis, and I_{XX} is the second moment of area of the lamina about some other, parallel, axis XX a distance h from NA, then the parallel axis theorem states

$$I_{XX} = I_{NA} + Ah^2$$

where A is the area of the lamina.

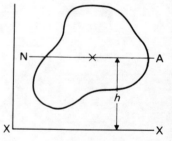

Applying the theorem to a rectangular lamina of breadth b and depth d, where XX lies along one edge.

$$I_{XX} = I_{NA} + Ah^2$$

$$= \frac{bd^3}{12} + (bd)\left(\frac{d}{2}\right)^2$$

$$= \frac{bd^3}{12} + \frac{bd^3}{4}$$

$$= \frac{bd^3 + 3bd^3}{12} = \frac{4bd^3}{12} = \frac{bd^3}{3}$$

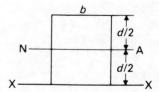

a result already determined previously from first principles.

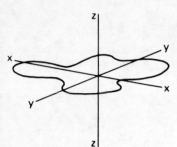

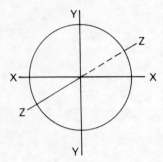

The perpendicular axis theorem

If I_{XX} and I_{YY} are the second moments of area of a lamina shape *in the same plane*, and at right-angles to each other, and I_{ZZ} is the second moment of area of the lamina perpendicular to the plane and passing through the intersection of I_{XX} and I_{YY}, the perpendicular axis theorem states:

$$I_{ZZ} = I_{XX} + I_{YY}$$

The second moment of area of a circular lamina about an axis through its centre in the plane of the lamina, i.e. about any diameter

Given a circular lamina of radius R, diameter D. I_{XX} and I_{YY} are the second moments of area of the lamina about two mutually perpendicular diameters XX and YY. I_{ZZ} is the second moment of area of the lamina about an axis through the centre of the lamina, but perpendicular to the plane of the lamina (the polar second moment of area $= \pi D^4/32$, see Appendix 3).

Using the perpendicular axis theorem,

$$I_{ZZ} = I_{XX} + I_{YY}$$

but, due to the complete similarity of the lamina $I_{XX} = I_{YY}$

$$\text{therefore } I_{ZZ} = 2I_{XX}$$

but $I_{ZZ} = \dfrac{\pi D^4}{32}$ (see Appendix 3)

$$\text{therefore } \frac{\pi D^4}{32} = 2I_{XX}$$

$$\text{therefore } I_{XX} = \frac{\pi D^4}{2 \times 32}$$

$$= \frac{\pi D^4}{64}$$

Thus the second moment of area of a circular lamina about any diameter is $\pi D^4/64$.

Appendix 3
The second moment of area of a lamina about an axis through its centroid at right-angles to the plane of the lamina (the polar second moment of area)

Given a plane lamina with an axis ZZ through its centroid and perpendicular to the plane of the lamina. Consider an elemental area δA, distance r from ZZ.

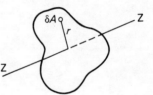

First moment of area of element about ZZ $= \delta A \times r$.

Second moment of area of element about ZZ $= \delta A r^2$.

Total second moment of area of whole lamina about ZZ is equal to the sum of the second moments of all such elements about ZZ $= \int r^2 \, dA$

   i.e. $I_{ZZ} = \int r^2 \, dA$

applying the theory to an actual shape.

The second moment of area of a circular lamina about an axis through its centre and at right-angles to the plane of the lamina

Given a circular lamina of radius R, diameter D, and an axis ZZ perpendicular to the plane of the lamina, and passing through the centre of the lamina. Consider an annular element, radius r and thickness δr. Since δr is very small, then area of element $\simeq 2\pi r \times \delta r$.

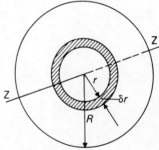

First moment of area of element about ZZ $= 2\pi r \times \delta r \times r$.

Second moment of area of element about ZZ $= 2\pi r \times \delta r \times r^2$
$$= 2\pi r^3 \times \delta r$$

Total second moment of area of all such elements from $r = 0$ to $r = R$ is

$$\int_0^R 2\pi r^3 \, dr = 2\pi \left[\frac{r^4}{4} \right]_0^R = \frac{\pi R^4}{2}$$

or, substituting

$$R = \frac{D}{2}, \; I_{ZZ} = \frac{\pi (D/2)^4}{2} = \frac{\pi D^4}{32}$$

I_{ZZ} is referred to as the **polar second moment of area** of the lamina, since there is a tendency for the lamina to rotate about the axis ZZ. In engineering the polar second moment of area is often given the symbol J rather than I, in order to distinguish clearly between them.

Thus $J = \dfrac{\pi D^4}{32}$ for a circular lamina.

Appendix 4
Moment of inertia

When considering solids rather than laminas the elemental areas become elemental masses, and the centroid becomes the centre of gravity. Confusingly the symbol I is still used for moment of inertia, and so it is important to view the content of problems and the context in which the symbol is used.

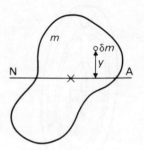

Given a mass m, with an axis NA passing through its centre of gravity. Consider an elemental mass δm at a distance y from NA.

First moment of mass about NA $= \delta my$.

Second moment of mass about NA $= \delta my^2$.

Total second moment of mass, or the moment of inertia, about NA $= \int y^2 \, dm$, i.e.

$$I = \int y^2 \, dm$$

In many engineering problems the moment of inertia of a body is required about an axis which the body is spinning.

Thus if ZZ is the required axis,

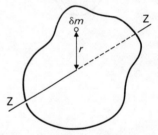

First moment of mass about ZZ $= \delta mr$.

Second moment of mass about ZZ $= \delta mr^2$.

Total moment of inertia is then $\int r^2 \, dm$.

The moment of inertia of a flat circular disc

Applying the theory to a flat disc of radius R and thickness t. Let m be the mass per unit volume. Consider an annular element of thickness δr and radius r.

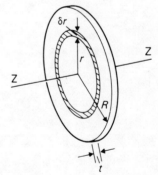

Mass of element $= 2\pi r \delta r t m$

First moment of mass about ZZ $= 2\pi r \delta r t m r$

Second moment of mass about ZZ $= 2\pi r \delta r t m r^2$

$\qquad\qquad\qquad\qquad\qquad\quad = 2\pi r^3 m t \delta r$

Total second moment of mass (moment of inertia) is equal to the sum of the second moments of all such elements from $r = 0$ to $r = R$

$$= \int_0^R 2\pi m t r^3 \, dr = 2\pi m t \left[\frac{r^4}{4}\right]_0^R$$

$$= \frac{2\pi m t R^4}{4} = \frac{\pi m t R^4}{2}$$

But $m\pi R^2 t = M$, the total mass of the disc.

Therefore, moment of inertia of disc about an axis through its centre perpendicular to plane of disc $= MR^2/2$.

Radius of gyration of a flat circular disc

From equation (3.10) $I = mk^2$

Therefore for a circular disc $k^2 = \dfrac{R^2}{2}$ or $k = \dfrac{R}{\sqrt{2}} = 0\cdot707R$

The moment of inertia of a flat annular ring

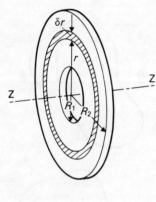

Given a flat annular ring of internal radius R_1 and external radius R_2 and thickness t. Let m be the mass per unit volume. Consider an annular element of thickness δr and radius r.

Mass of element $= 2\pi r \delta r t m$
First moment of mass about ZZ $= 2\pi r \delta r t m r$
Second moment of mass about ZZ $= 2\pi r \delta r t m r^2$
$= 2\pi r^2 m t \delta r$

Total second moment of mass (moment of inertia) is equal to the sum of the second moments of all such elements from $r = R_1$ to $r = R_2$, i.e.

$$I_{ZZ} = \int_{R_1}^{R_2} 2\pi r^3 mt \, dr = 2\pi mt \left[\frac{r^4}{4}\right]_{R_1}^{R_2} = \frac{\pi mt}{2}(R_2^4 - R_1^4)$$

$$= \frac{\pi mt}{2}(R_2^2 - R_1^2)(R_2^2 + R_1^2)$$

but $\pi(R_2^2 - R_1^2)mt$ is equal to the mass of the whole ring. Therefore moment of inertia of an annular ring about an axis perpendicular to its plane is

$$\frac{M(R_2^2 + R_1^2)}{2}$$

Answers to Problems

1.1
1. 330 N
2. 10.2 m/s; 3.84 m/s; 7.02 kN
3. 42.1 N
4. 1.367 m/s to right; 15 m
5. 784.8 m/s; 92.4 kN
6. 0.7 m/s; 14.7 mm; 0.04255 s
7. 10 Mg; 1.82 m/s; 5 Mg; 1.923 m/s
8. 10.2 m/s; 1.675 m/s; 12.53 kN
9. 2 m/s; 0.001 s
10. 7.08 m/s; 20.64 kN

1.2
1. 408.7 kJ/min; 291.7 kJ/min
2. 1435.5 kJ/s; 114.45 kJ/s
3. 3200 kJ
4. 37016 kJ/min
5. 122 kJ; 48 800 kJ/min
6. 1200 J
7. 80.5 kJ; 115 N
8. 1.49 kJ; 38.6 m/s
9. 208.5 kJ; 131.8 kJ

1.3
1. 98.1 J; 4.43 m/s; 68.67 J
2. 3.26 m
3. 22.15 m/s
4. 835 kJ; 4.2 kN
5. 54 J; 0
6. 2.3 m/s
7. 8.77 m/s; 769 m
8. 38.7 m; 304 m
9. 15.9 m; 4.77 s
10. 1620 m; 3 min
11. 131.7 N; 6.7 s
12. 600 N
13. 180 J; 1.93 m/s
14. 356.7 kJ; 76.73 kJ; 1.46 m; 52.56 kN
15. 294.3 J; 4.43 m/s; 189.2 J

1.4
1. 136 kW
2. 622 kW; 20 MW
3. 74.4 kW

1.5
1. 1.256 kJ
2. 4.19 kW
3. 0.72 rev/s
4. 6.0 joules
5. 750 joules
6. 4.46 kJ
7. 61.31 kJ
8. 2.72 kW; 432.5 Nm; 13.6 kW
9. 57.5 W
10. 1018.6 N; 203.7 Nm
11. 0.4 kNm; 6.28 kW
12. 2.45 kW; 1.056 kW

2.1
1. 0.98 mm
2. 35.36 mm; 0.148 mm
3. 35.7; 62.6; 142.5 MN/m^2; 0.114 mm
4. 17.21 MN/m^2; 103.26 MN/m^2
5. 221 mm
6. 92.9 MN/m^2; 43.3 MN/m^2; 0.068 mm
7. 2100 mm^2; 89.4 MN/m^2; 35.7 MN/m^2
8. 15.96 MN/m^2; 7.36 MN/m^2; 0.0116 mm

2.2
1. $\sigma_B = 138.75$ MN/m^2; $\sigma_S = 118.2$ MN/m^2
2. $\sigma_C = 116.8$ MN/m^2; $\sigma_s = 42.05$ MN/m^2
3. 12.6 kN
4. 30.8 MN/m^2
5. 110 MN/m^2

2.3
1. 0.8 mm
2. 268.8 kN
3. 91.5 mm
4. 177 N/mm^2; 306 N/mm^2
5. 9.91 mm
6. 0.332 m
7. 345.9 N/mm^2
8. 26.23 mm; 432 N/mm^2

2.4
1 0.001107 mm
2 0.0000231
3 29.4 mm^3 decrease
4 0.32
5 1.11 mm^2
6 4.482 mm^2 increase
7 10.9 mm^2 increase
8 4.2 mm^3 increase
9 $\epsilon_a = 0.625 \times 10^{-3}$;
   $\epsilon_c = 1.11 \times 10^{-3}$

2.5
1 118 MN/m^2; 490.7 m
2 ii is 16 $\times$ stronger than i;
   9.5 kN
3 3.46 mm; 0.943 Nm
4 3.3 kN/m; 700.9 m
5 4.133 kNm; 4.9 kN
6 1.1 mm

2.6
1 356 $\times$ 171
2 838 $\times$ 292
3 610 $\times$ 229

2.7
1a 211.1 kN: 288.9 kN
 b 4.27 kN; 4.93 kN
 c 4.67 $\times$ 10^4 kN; 5.33 $\times$ 10^4 kN
 d 206$\frac{2}{3}$ kN; 253$\frac{1}{3}$ kN
 e 155.56 kN; 344.44 kN
 f 1.93 $\times$ 10^5 N; 6.07 $\times$ 10^5 N
 g 806$\frac{2}{3}$ N; 913$\frac{1}{3}$ N
 h 500 kN; 2000 kN

2a R = 100 kN; M = 740 kNm
 b R = 5.7 kN; M = 39.6 kNm
 c R = 1750 kN; M = 7625 kNm
 d R = 140 N; M = 765 Nm

2.8
1a

b

c

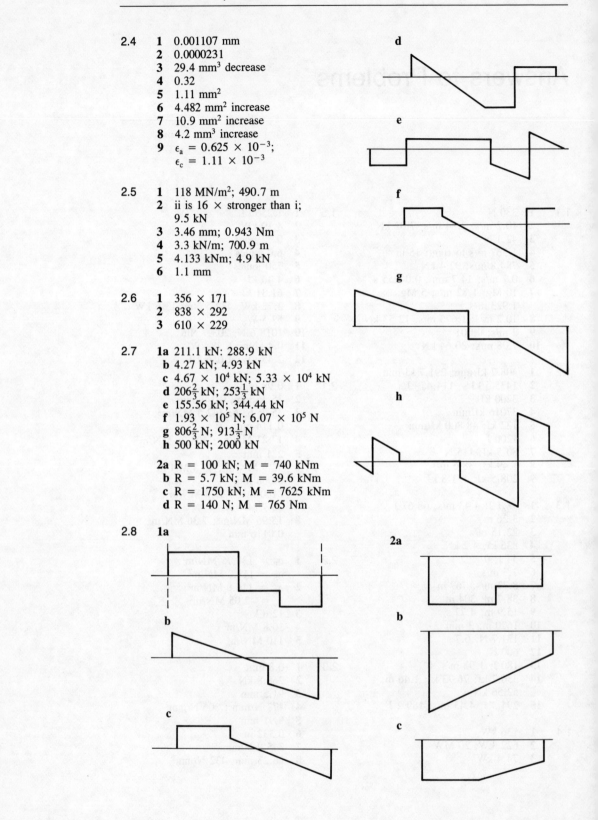

d

e

f

g

h

2a

b

c

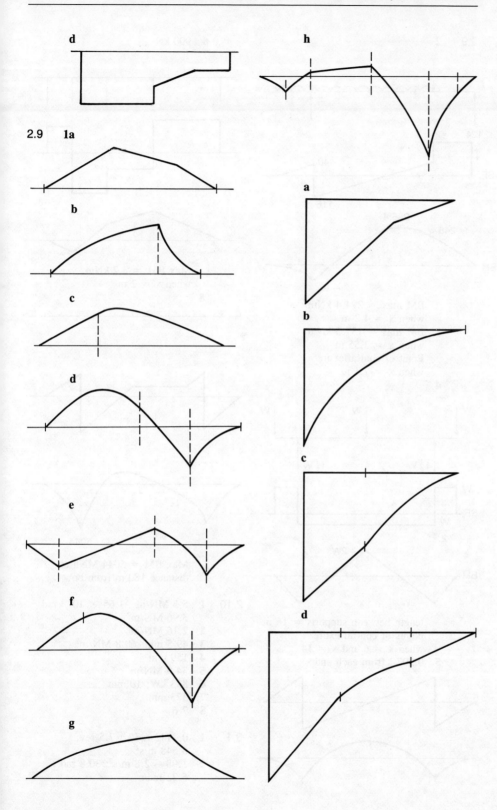

2.9   1a

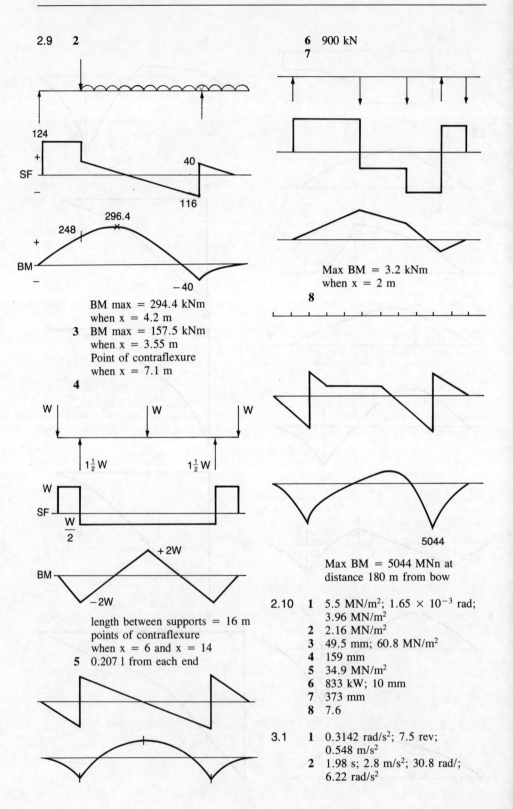

2.9   **2**

124

SF   +

40

−

116

296.4

248   +

BM

−

−40

BM max = 294.4 kNm
when x = 4.2 m

3   BM max = 157.5 kNm
when x = 3.55 m
Point of contraflexure
when x = 7.1 m

4

W   W   W

$1\frac{1}{2}$ W $1\frac{1}{2}$ W

W
SF
$\frac{W}{2}$

+ 2W

BM

−2W

length between supports = 16 m
points of contraflexure
when x = 6 and x = 14

5   0.207 l from each end

6   900 kN

7

Max BM = 3.2 kNm
when x = 2 m

8

5044

Max BM = 5044 MNn at
distance 180 m from bow

2.10   **1**   5.5 MN/m^2; 1.65 × 10^{-3} rad;
        3.96 MN/m^2
     2   2.16 MN/m^2
     3   49.5 mm; 60.8 MN/m^2
     4   159 mm
     5   34.9 MN/m^2
     6   833 kW; 10 mm
     7   373 mm
     8   7.6

3.1   **1**   0.3142 rad/s^2; 7.5 rev;
        0.548 m/s^2
     2   1.98 s; 2.8 m/s^2; 30.8 rad/;
        6.22 rad/s^2

3   0.11 m/s^2; 0.366 rad/s^2; 8.5 m; 4.51

4   0.1885 m/s; 0.3495 rad/s^2; 0.0314 m/s^2; 1.698 m

5   4.57 rad/s^2; 22.85 rad/s; 9.09; 4.0 m/s

6   2 m/s^2; 26.67 rad/s^2; 255 rev/min; 2.123 rev

7   1.3 m/s; 0.65 m/s^2; 13 rad/s^2; 4.14 rev

8   2.626 rev; 132 rad/s^2; 37.5 m/s; 9.48 m

3.2
1. 145.9 rev/min
2. 405; 2.35 kNm; 6.27 kN
3. 960 N; 1.0 rad
4. 6.6 kNm
5. 1.66 m/s^2
6. 0.57 m/s^2; 15.57 kN; 9532 Nm
7. 1.31 m/s^2; 111.2 N; 68 N
8. 402 Nm; 5693 Nm

3.3
1. 84.29 N; 104.5 N; 38.6 rev/min
2. 8.8 rev/min
3. 98.1 N; 134 rev/min
4. 21.1 kN; 23.07 kN
5. 1778 N
6. 51.4 km/h
7. 723 mm
8. overturns at 22.25 rev/min
9. 84.4 m
10. 276 rev/min; 1184 N; 230 Nm
11. 45.6 kN; 32.85 kN; 124 km/h
12. 81.25 knots; 20.5 kN
13. 28.7 kN; 28.7 kN; 69.4 kN; 69.4 kN; 41.5 m
14. 82.5 km/h; 171 km/h
15. 9.95°
16. 9.06 kN

3.4
1. 25.66 N; 0.795 m/s; 1.1 s
2. 0.26 m; 0.1216 m
3. 0.036 s; 379 N
4. 0.477 m; 0.274 m; 0.1 s
5. 49.35 N; 0.073 m; 1.11 m/s
6. 0.45 s; 4.9 m/s^2
7. 2500 N/m; 46.5 mm
8. 191.9 N/m; 48.3 mm
9. 6.3 kg; 1.584 m/s^2; 71.8 N
10. 261 N/m; 75 mm
11. 2.32 s; 1.335 m; 3.13 m/s; 4.9 m/s^2
12. gain 19 s

3.5
1. 25 rev/min; 16
2. 282.66 rev/min; 6.55 Nm

3   31 rev/min

4   366.5 Nm; 205.7 rev/min

5   6995 rev/min

6   328.2 kJ; 820 N

7   105.6 kN/m

8   86 cm

4.1
1. 22.2 MN/m^2; 11.1 MN/m^2
2. 16 bar
3. 85 mm
4. 1.00176 d; 1.00041 l; 1.00747 V
5. 5.43 bar; 0.0122 m^3
6. $\sigma_S = 3.08$ MN/m^2; $\sigma_A = 4.93$ MN/m^2
7. $\sigma_S = 110$ MN/m^2; $\sigma_C = 193.5$ MN/m^2

4.2
1. $\sigma = 72$ MN/m^2; $\tau = 13$ MN/m^2
2. $\sigma = 25.1$ MN/m^2; $\tau = 38.8$ MN/m^2
3. $\sigma = 7.8$ GN/m^2; $\tau = 4.75$ GN/m^2
4. $\sigma_1 = 95.2$ MN/m^2; $\sigma_2 = -50.2$ MN/m^2; 24.6° and 114.6°; 72.7 MN/m^2
5. $\sigma = 67$ MN/m^2; $\tau = 22$ MN/m^2; Resultant = 70.5 MN/m^2
6. $\tau = 49$ MN/m^3; $\tau_{max} = 70$ MN/m^2
7. $\sigma_1 = 102.8$ MN/m^2; $\sigma_2 = 2$ MN/m^2; 31.5° and 121.5° clockwise from plane carrying hoop stress; $\tau_{max} = 50.4$ MN/m^2
8. 0.155 m; 0.183 m
9. 0.528×10^{-3}

5.1
1a. 3.474×10^{-5} m^4
 b. 3.584×10^{-5} m^4
2a. 8.015×10^{-6} m^4
 b. 1.20655×10^{-4} m^4
3a. 7.05×10^{-7} m^4
 b. 4.41×10^{-7} m^4
4a. 3.75×10^{-5} m^4
 b. 1.36×10^{-5} m^4

5.2
1. +16 MN/m^2; −176 MN/m^2
2. 32.7 mm
3. 326 N
4. 37.86 MN/m^2
5. −8 MN/m^2; −23.7 MN/m^2; −8 MN/m^2; +7.66 MN/m^2

5.3
1. −12.2 MN/m^2; +6.42 MN/m^2
2. 1.24 m

3 -363 MN/m^2

6.1   **1**   0.4 joules
2   0.423 joules
3   166 J
4   7.67 kN/m^2
5   279 MN/m^2

6.2   **1a**   0.132 joules
b 2.62×10^{-5} joules
c   4.15 kJ
d   0.066 joules

6.3   **1a**   100 joules
b   0.44 joules
c   2030 kJ
d   23 J

6.4   **1**   3.77 J; 0.665°
2   15.4 mm
3   119 mm

6.5   **1**   1.29 MN/m^2
2   12 mm; 0.83 m
3   104.9 MN/m^2
4   38.06 MN/m^2
5   151 mm
6   3623 cm^3

7.1   **1**   12.7 mm
2   0.00613 rad; 9.2 mm
3   0.00469 rad; 7.6 mm
4   0.6 N

5 $\pm \dfrac{5 \times 10^4}{EI}$ rad;

$-\dfrac{7.5 \times 10^4}{EI}$ m at each load

$0;\ -\dfrac{10^5}{EI}$ at mid-span

7.2   **1**   0.00052 rad; 0.344 mm
2   0.0146 rad; 60.2 mm
3 9×10^{-5} rad; 0.036 mm;
0.072 mm
4   2.83 mm
5   21 mm

7.3   **1**   5.65 mm; 12.9 mm; 10.7 mm
2   0.00017 rads; 0.371 mm
3   10.9 mm at 4.16 mm from A
4   35.6 mm at C
5 -0.019 rad; $+18.24$ mm;
-23.55 mm

6 -19.45 mm when x = 3.92 m
7 -16.02 mm when x = 4.46 m

8.1   **1**   274.3 watts; 368 watts
2   1221 N; 10.69 kW
3   3
4   4.06 kW
5   45.6 watts

8.2   **1**   3.04 kW
2   8.19 kW; 22.36 m/s
3   13.3 kW; 30.15 m/s
4   490.9 kW; 50.64 m/s
5   45.49 kW; 25.4 m/s
6   18 kN; 14.2 kN
7   503 watts
8   7

8.3   **1**   3898 N
2   6.2 kW
3   1 kW less
4   8
5   3074.9 N; 61.8 mm
6   3.36 s; 2.16 rad/s; 8.62 kJ

8.4   **1**   10.97 watts
2   78.38 watts
3   109.9 watts
4   5.19 kW

8.5   **1**   34 W
2   790 W
3   21.5 W
4   37.7 N

9.1   **1**   1.5 m/s; 12.2 rad/s
2   5.6 m/s; 11 rad/s
3   3.55 m/s; 2.6 m/s; 7.1 rad/s
4   3.5 m/s; 13.1 rad/s
5   1.7 m/s; 12.4 rad/s
6   1.2 m/s; 12 rad/s
7   1.38 m/s; 7.5 rad/s
8   1.2 rad/s; 0.26 m/s

9.2   **1**   166.6 kg
2   8 kJ; 0.052
3   59.3 Nm; 1.55 kJ; 17.6 kg m^2
4   235 rev/min; 678

9.3   **1**   1.026 kg at 0.5 m and 180°
from 60 kg mass
0.426 kg at 1.2 m and 0° from
60 kg mass
2   74 N at 0.5 m; 22.2 N at 1.2 m
3   84 kg at 0.3 m; 36 kg at 0.7 m
both 180° displaced from 75 kg
mass

4   both 44.4 kg at 22° 37′ either
    side of an axis of symmetry
5   216° 11′ between A and B;
    297° 55′ between A and D;
    0.697 m; 7.57 kg
6   A is 274°, B is 151° from
    balance weight
    59.2 N at bearing closest to A
    98.7 N at bearing 0.75 m from
    A
7   26 kg midway between A and B,
    at 193° to A
    18 kg midway between B and C,
    at 310° to A

10.1   1   25.65 N; 0.795 m/s; 1.1 s
       2   0.036 s; 379 N

3   0.28 s; 0.134 m/s; 4 m/s^2;
    0.029 s
4   0.145 s
5   1.59 Hz; −0.02 m;
    −0.335 m/s; −2 m/s^2
6   0.89 Hz; 0.874 Hz
7   0.4 s
8   0.313 m
9   91.2 kg
10 63.5 Hz
11 0.324 kg m^2
12 0.67 m
13 5.24 Hz
14 3.74 Hz
15 2.95 Hz
16 11.14 kN each
17 945 rev/min; 846 rev/min
18 1497 rev/min; −0.0455 mm

Index